Ecosystem Ecology

A New Synthesis

What can ecological science contribute to the sustainable management and conservation of the natural systems that underpin human well-being?

Bridging the natural, physical and social sciences, this book shows how ecosystem ecology can inform the ecosystem services approach to environmental management. The authors recognise that ecosystems are rich in linkages of varying strength between biophysical and social elements that generate powerful intrinsic dynamics. Unlike traditional reductionist approaches, the holistic perspective adopted here is able to explain the increasing range of scientific studies that have highlighted unexpected consequences of human activity, such as the lack of recovery of cod populations on the Grand Banks despite nearly two decades of fishery closures, or the degradation of Australia's fertile land through salt intrusion.

Written primarily for researchers and graduate students in ecology and environmental management, it provides an accessible discussion of some of the most important aspects of ecosystem ecology and the potential relationships between them.

David G. Raffaelli is Director of the UK's NERC Centre, UKPopNet. He has written extensively on aspects of ecosystem ecology, especially food webs and integrated catchment ecology, and more recently has become extensively involved with the ecosystem services approach to the management of natural resources within the UK and Europe through his work with DIVERSITAS, UKBRAG, the Royal Society's Global Environment Research Committee, Defra and the British Ecological Society (BES).

Christopher L. J. Frid is Professor of Environmental Science and Marine Biology at the University of Liverpool and a long-standing member of the BES. He is a member of Defra's Marine Fisheries Science Advisory Group and the Council of the Marine Biological Association of the United Kingdom. He has written extensively on aspects of marine ecology and human impacts on marine ecosystems and has been a major proponent of the development of the ecosystem approach to marine management.

Ecological Reviews

Ecological Reviews will publish books at the cutting edge of modern ecology, providing a forum for volumes that discuss topics that are focal points of current activity and likely long-term importance to the progress of the field. The series will be an invaluable source of ideas and inspiration for ecologists at all levels from graduate students to more-established researchers and professionals. The series will be developed jointly by the British Ecological Society and Cambridge University Press and will encompass the Society's Symposia as appropriate.

Biotic Interactions in the Tropics: Their Role in the Maintenance of Species Diversity
Edited by David F. R. P. Burslem, Michelle A. Pinard and Sue E. Hartley

Biological Diversity and Function in Soils
Edited by Richard Bardgett, Michael Usher and David Hopkins

Island Colonization: The Origin and Development of Island Communities
By Ian Thornton
Edited by Tim New

Scaling Biodiversity
Edited by David Storch, Pablo Margnet and James Brown

Body Size: The Structure and Function of Aquatic Ecosystems
Edited by Alan G. Hildrew, David G. Raffaelli and Ronni Edmonds-Brown

Speciation and Patterns of Diversity
Edited by Roger Butlin, Jon Bridle and Dolph Schluter

Ecology of Industrial Pollution
Edited by Lesley C. Batty and Kevin B. Hallberg

Ecosystem Ecology

A New Synthesis

Edited by

DAVID G. RAFFAELLI
Environment, University of York, York, UK

CHRISTOPHER L. J. FRID
School of Environmental Sciences, University of Liverpool, Liverpool, UK

CAMBRIDGE UNIVERSITY PRESS
Cambridge, New York, Melbourne, Madrid, Cape Town, Singapore,
São Paulo, Delhi, Dubai, Tokyo

Cambridge University Press
The Edinburgh Building, Cambridge CB2 8RU, UK

Published in the United States of America by Cambridge University Press, New York

www.cambridge.org
Information on this title: www.cambridge.org/9780521513494

First published 2010

Printed in the United Kingdom at the University Press, Cambridge

A catalogue record for this publication is available from the British Library

Library of Congress Cataloguing in Publication data
Ecosystem ecology : a new synthesis / [edited by] David G. Raffaelli, Christopher L. J. Frid.
p. cm. – (Ecological reviews)
ISBN 978-0-521-51349-4 (hardback)
1. Biotic communities–Research. 2. Human ecology–Research. 3. Ecosystem management–Research. I. Raffaelli, D. G. (Dave G.) II. Frid, Chris. III. Title. IV. Series.
QH541.2.E256 2010
577–dc22

ISBN 978-0-521-51349-4 Hardback
ISBN 978-0-521-73503-2 Paperback

Contents

Contributors

ANDY FENTON
School of Biological Sciences, University of Liverpool, Liverpool, UK

CHRISTOPHER L. J. FRID
School of Environmental Sciences, University of Liverpool, Liverpool, UK

ROY HAINES-YOUNG
Centre for Environmental Management (CEM), School of Geography, University of Nottingham, Nottingham, UK

KEVIN EDSON JONES
Management School, University of Liverpool, Liverpool, UK

ODETTE A. L. PARAMOR
School of Environmental Sciences, University of Liverpool, Liverpool, UK

MARION POTSCHIN
Centre for Environmental Management (CEM), School of Geography, University of Nottingham, Nottingham, UK

DAVID G. RAFFAELLI
Environment, University of York, York, UK

ANNA R. RENWICK
College of Life Science and Medicine, University of Aberdeen, Aberdeen, UK

JAMES C. R. SMART
Environment, University of York, York, UK

MATTHEW SPENCER
School of Environmental Sciences, University of Liverpool, Liverpool, UK

PAUL C. STOY
School of GeoSciences, University of Edinburgh, Edinburgh, UK

PIRAN C. L. WHITE
Environment, University of York, York, UK

Preface

What can ecological science contribute to the sustainable management and conservation of the natural systems that underpin human well-being? This is a question that is taxing many professional ecologists, learned societies and science funders. The question has been driven by both the increased awareness of the present ecological crisis and the publication of several documents influential at the highest political level, such as the Intergovernmental Panel on Climate Change (IPCC) reports, the Stern Review, GEO4 and, most relevant to the present volume, the Millennium Ecosystem Assessment. The impetus and stimulation for this volume came in part from workshops hosted by UK Pop Net and the British Ecological Society (BES) in 2003 and 2007, respectively, which aimed to seek an answer to the question: how can mainstream ecology, and by definition the ecologists within learned societies like the BES, contribute to national and international initiatives aimed at implementing a holistic ecosystem approach for environmental management? That workshop revealed a huge potential within the community but also frustrations about, and ignorance of, the different perspectives on ecosystem ecology held by the different sectors within mainstream ecology: reductionist versus holistic approaches, inter-disciplinary versus mono-disciplinary approaches, those which recognise humans as part of versus apart from the ecosystem.

Ecosystem Ecology implies both a different perspective and a different approach to the science. The more holistic view tends to regard the ecosystem as rich in ecological linkages, some of which may be strong but many of which will be individually weak. However, the number of linkages provides a system with a powerful intrinsic dynamic. It therefore follows that a reductionist approach to the study of the system may readily identify any strong links but may fail to correctly understand the system's topology and dynamics. Legislative and environmental management frameworks have in recent years placed greater emphasis on a holistic, ecosystem approach. In part this might be a response to political power passing to the 1960's 'silent spring' generation but it might also be due to the increasing range of scientific studies that have highlighted unexpected, based on reductionist views, consequences of human activities.

Examples include the lack of recovery of cod populations on the Grand Banks after nearly two decades of fishery closures, the massive underestimation of the importance of mature forests to carbon sequestration and the impacts of an alien (non-native) species of small jellyfish on the ecology of the Black Sea.

This volume is not an attempt to provide an overarching theory or framework that will bring these different approaches under a single banner. Rather it aims to make accessible, for those willing to make the journey, approaches which might otherwise seem too demanding or even not worthwhile to tackle at first sight. Many of the aspects of ecosystem ecology that are explored in this book have been around and actively pursued for some time, but often without explicit acknowledgement of the potential connections, relationships and synergies between them.

In Chapter 1, we briefly review some of these different approaches and attempt to place their origins in an historical context in order to account for their often divergent trajectories and isolation of the different research schools, and we illustrate the potential linkages and analogies between them to encourage better integration of those ideas. Chapter 2 examines theoretical approaches at the population, assemblage and ecosystem scales and the connections and links between them. This raises the question as to whether increasing computer power and hence the ability to run more complex models has now moved to the point where our focus should return to the collection and analysis of empirical data on the systems of interest. The linking of the physical world, as constrained by the Laws of Thermodynamics, with the response of the biological part of the ecosystem forms the central theme of Chapter 3. These linkages illustrate the dynamic nature of ecosystems and the need to study them from this perspective if we are to develop the understanding necessary to then develop environmental management schemes.

As environmental management has moved up the political agenda, science has been asked to provide measures of the health of the environment. Chapter 4 examines the concept of ecosystem health and the approaches available to assess it. With politicians trying to balance the need to deliver all 'three pillars' of sustainability (ecological, social and economic), so ecosystem health assessments often feature measures of the human aspects of the ecosystem. Chapter 5 examines how interdisciplinary studies of the ecosystem are developing and the barriers that are being encountered as social scientists, economists and ecologists attempt to bring their expertise to bear simultaneously on problems of sustainable ecosystem management.

In Chapter 6 we further explore the science that links ecosystem processes with human activities, the ecosystem services approach. Once again this highlights the benefits of holistic ecology and the need for interdisciplinary working. In Chapter 7 we draw together many of the themes developed in earlier chapters and consider explicitly how ecosystem ecology is relevant to those who make and implement environmental, in its broadest sense, policy.

Finally, we wish to thank the team of authors for agreeing to be part of this project, for contributing their expertise and for their patience and forbearance as we have struggled to pull the whole together. In the best traditions of ecosystem science we hope that the whole is more than the sum of the parts!

CHAPTER ONE

The evolution of ecosystem ecology

DAVID G. RAFFAELLI
Environment, University of York
CHRISTOPHER L. J. FRID
School of Environmental Sciences, University of Liverpool

Introduction

The sustainable use, management and conservation of ecosystems, as promoted by the Convention on Biological Diversity's Ecosystem Approach (United Nations 1992), and recent initiatives such as the Millennium Ecosystem Assessment (United Nations 2005), emphasise the inter-dependence between ecological systems and human well-being. Healthy social systems demand healthy ecosystems and vice versa. This emergent world view is compelling and persuasive to conservationists, policy makers and managers alike, because it implies win-win solutions for nature conservation and for human development which relies on the continued provision of ecosystem goods and services. Ecosystem management within this context requires a holistic approach that acknowledges the need to work with and across a broad range of natural, physical, social and economic sciences. Whilst there are many successful programmes which have achieved this, mainstream ecologists who have so much to bring to the table have been slow to embrace such approaches. Jones and Paramor (this volume) consider many of the important cultural challenges. The view that humans are part of, not apart from, the biophysical system in which they are embedded has not always sat comfortably with academic researchers, who have traditionally seen their prime focus on, and responsibility to, either the natural system or to broader societal goals, but rarely both. In addition, there are misunderstandings and fears about what holistic ecosystem approaches really are, in turn due to the divergent pathways along which different sections of the ecological community have developed. These issues are not new: the tension between reductionist and holistic approaches has bedevilled the development of a coherent discipline of ecosystem ecology and issues of working across the disciplines recur throughout the short history of ecology.

The aim of this chapter is to describe some of that rocky landscape through which ecosystem ecology and its research community have travelled over the past sixty years or so, from the problems of defining what an ecosystem

Ecosystem Ecology: A New Synthesis, eds. David G. Raffaelli and Christopher L. J. Frid. Published by Cambridge University Press.

actually is in the 1930s, with the only too familiar issues of loose terminology and the all-things-to-all-people concept of an ecosystem. We then provide a retrospective analysis of the most ambitious international ecosystem research programme ever mounted, the International Biological Programme (IBP) of the 1960s and 1970s, an initiative that laid the foundations of ecosystem ecology. We discuss the strengths and weaknesses of the IBP, which have a bearing on how ecosystem ecology might develop in the future. The quantitative holistic approach of systems analysis which underpinned much of the IBP has never achieved the prominence and potential it should have enjoyed and we explore the reasons for this. We then move on to the new emerging frameworks and concepts within Resilience Theory to discuss the potential of this area for ecosystem science and in particular its implications for management. Finally, we reflect on what we can learn from the history of these aspects of the development of ecosystem research so that future endeavours do not result in the same mistakes or ignore the hard lessons learned.

Origins of the concept of the ecosystem

The emerging holistic view of humans and their environment is hardly a novel one: it is fundamental to the human condition and articulated in the articles of faith of many of the world's religions that recognise the inter-connectedness of natural, physico-chemical and human dimensions of the environment. However, the formalisation of the concept of natural ecosystems in a scientific sense began in the early part of the twentieth century, chiefly with the perspectives of Clements and Tansley (for an excellent historical review, see Sheail 1987). Both Clements and Tansley were plant ecologists and their perspectives on natural systems were markedly influenced by the vegetation successional patterns they witnessed around them, although in quite different ways. Clements held that plant communities could be viewed as super-organisms with different developmental stages having their own organic unity. Whether Clements came to this view through his empirical observation of nature (views formed mainly in the environment of the mid west of the US) or whether this perspective was an a-priori concept later supported by empirical observation is difficult to discern at this point in history, given the continual cross-informing of theory and observation in research which all researchers experience. Other leading ecologists of the time, notably Tansley and Gleason (informed mainly by experience of the New England landscape), became increasingly doubtful of this Clementsian world view, taking a more individual-based, reductionist approach, and seeing the patterns in plant communities which develop over time as inevitable expressions of the interactions between individual species, a view that prevails in mainstream ecology to this day. Gleason seems to have suffered greatly for taking what many today would consider a reasonable and sensible stance, becoming one of the first of a long line of 'ecological outlaws' (Sheail 1987), whereas Tansley's status and reputation were seemingly unassailable in this respect.

The 'super-organism' and 'emergent pattern' (broadly equivalent to a holistic versus reductionist) debate took on an uncompromising tone in later years, although Tansley's commentaries and remarks show him to have been surprisingly pluralistic in many respects. He acknowledged that 'the strength of the Clementsian system lay in its philosophical sweep and comprehensiveness' (Sheail 1987, p. 61), holding that ecological concepts were 'creations of the human mind which we impose on the facts of nature' (Tansley 1914, from Sheail 1987, p. 60). In other words, ecological concepts are heuristic devices or semi-abstract models which help to drive the field forward as these devices are explored to their limits, evolve or are overturned (Sheail 1987, p. 63). The Clementsian–Gleason–Tansley debate, which must have seemed bitter at times, is highly relevant in the present context not only because of the outcome of that debate, but also because similar highly charged exchanges, in part based on misunderstandings concerning heuristic devices, occur today, exemplified in the present volume by the schism between reductionist and holistic approaches to ecosystem ecology.

Much of the difficulty in reaching a synthesis towards a unified approach to ecosystem ecology lies in the all-inclusiveness of the term 'ecosystem' (Willis 1997, Jax 2007). Whilst the basic concept has existed in many guises for at least a hundred years, the term itself was used in the 1930s by the British ecologist Roy Clapham and then refined by Tansley in an attempt to impose some rigour and consistency in a rapidly expanding discipline (Willis 1997). Tansley's definition was broad:

> the whole system (in the sense of physics), including not only the organism-complex, but also the whole complex of physical factors forming what we call the environment of the biome ... It is the systems so formed which from the point of view of the ecologist are the basic units of nature on the face of the Earth...These ecosystems, as we may call them, are of the most various kinds and sizes. They form one category of the multitudinous physical systems of the universe, which range from the universe as a whole down to the scale of the atom (Tansley 1935, from Lindeman 1942).

This view of the coupling of the biological and physical-chemical processes to form a single 'ecological system' seemed commonsensical to Tansley and his peers, as it does to most ecologists today. However, today we have additional evidence of the reality of this coupling through the emergent properties of ecosystems, specifically the congruence of the scaling of biological and physical processes in both terrestrial and marine systems. In both environments the rate of change in scale of temporal and spatial dynamics follows the same relationship, i.e. they align on the same slope. Interestingly, marine ecological systems are congruent with the underlying physical scaling, while in terrestrial systems the physical systems operate in a more dynamic manner and the biological responses are slower at each spatial scale. This simple analysis illustrates the close coupling of biological and physical dynamics, and hence the wisdom of an ecosystem concept that accommodates both, but also highlights fundamental differences in the dynamics of different types of ecosystem.

Since then, the term ecosystem has been conveniently co-opted for a variety of purposes, often to the dismay of those who fear that such looseness reduces the rigour of the science (see commentary by Sheail 1987, pp. 256–7). Most recently, Willis (1997) has offered the following definition: 'a unit comprising a community (or communities) of organisms and their physical and chemical environment, at any scale desirably specified, in which there are continuous fluxes of matter and energy in an interactive open system'. Willis suggests that the value of such a broad and all-inclusive definition is that the term provides a useful framework for predictive studies, rather than constructing boundaries around an exclusive discipline.

The present-day usage of the term 'ecosystem' within initiatives such as the Millennium Ecosystem Assessment (United Nations 2005) and the Ecosystems Approach (United Nations 1992) embraces a much greater swathe of environmental and social science than originally implied by any of the definitions described above (see also Jax 2007, and Haines-Young and Potschin, this volume). In particular, there has been a shift in the view of an ecosystem to one where people are considered part of an interactive holistic system, as opposed to humans being external drivers of change. Interestingly, Tansley's writings suggest that he would probably have welcomed the broadening of the concept to include human behaviour and the social sciences. Not only does he appear to have been remarkably tolerant of abstractions of nature (exemplified by his tolerance to Clements' heurisms), as long as they remained useful models for taking the field forward and were not taken past their logical limits, but he also lived and worked in a part of the world (the UK) where the profound influence of human activity and the way humans had shaped the landscape and its vegetation over several thousand years was taken as read, unlike the situation for many ecologists based in the New World.

Holistic frameworks for exploring complex, interacting systems: the contributions of Lindeman and Elton

Ecologists have long acknowledged the awesome complexity of the interacting systems with which they have to deal, and that if commonalities of process and pattern across different ecosystems are to be identified in a search for underlying 'laws', then ways of handling this complexity need to be found. At around the middle of the last century, new approaches to tackling this complexity were developing, the most notable of which were Elton's and Lindeman's frameworks (Elton 1927, Lindeman 1942). Raymond Lindeman's seminal paper is breathtaking in its scope and contribution. Published posthumously immediately after the author's tragically early death, the paper provided what has turned out to be an enduring framework that allowed, for the first time, plant and animal communities to be considered together, and which accommodated decomposers and non-living components. By grouping individual species into functional trophic types (primary, secondary, tertiary etc., producers and consumers), Lindeman provided a holistic scheme of considerably reduced complexity compared to the spider-web diagrams of food webs. He developed Elton's earlier descriptions of hierarchies of numbers and body sizes in animal food webs by describing pyramids of biomasses and flows of energy between functional trophic types (trophic levels) that could accommodate all types of organisation. This approach led to explorations of trophic-energy relationships and concepts such as ecological efficiencies which accounted for the limits to food-chain length previously observed by Elton, as well as allowing intriguing observations on the populations of 'vegetarian Chinese' compared to the 'more carnivorous English' that can be supported by a given level of production!

The International Biological Programme

The framework developed by Lindeman was a major step in the development of ecosystem science. It also provided the basis for much of the science that underpinned one of the most imaginative international programmes on ecosystems ever embarked upon: the International Biological Programme (IBP). Whilst little known or appreciated by today's generation of ecologists, this programme established many of the fundamental techniques and approaches that we now take for granted in ecosystem ecology. In addition, the IBP can be seen as the forerunner of those programmes and initiatives which are the focus of the present volume, such as the Millennium Ecosystem Assessment. The IBP was a sequel to the International Geophysical Year (1957–8) and, it has been claimed, partly a response to the rise of the molecular sciences in the 1950s and 1960s which 'posed a strong challenge both in academic status and financial support to the long-established macrobiological sciences and their concern with whole organisms and communities' (Collins and Weiner 1977). The gestation of the IBP is also associated with a recognition following World War II of the need to feed a growing world, particularly in developing countries, a need which in turn demanded a clear scientific understanding of the functioning of ecological systems and the limits to their production (Worthington 1965, 1975, 1983). An ambitious series of site-specific studies was established across the world, covering a great diversity of ecosystem types in over fifty countries. Each explored aspects of the fundamental basis of ecosystem productivity and human adaptation to those systems. Potentially, the programme was truly international, truly interdisciplinary and truly holistic.

The long-term beneficiaries of those research programmes include the editors of this volume and many other ecologists. Most importantly, a systems-analysis approach characterised the research programmes, facilitating comparisons and the search for commonalities between different ecosystem types (Figure 1.1). Many synthesis volumes and other publications have resulted from the IBP but the programme had a finite life (1964–74) and there probably remains much meta-analysis of the outcomes to be completed, even today. Underpinning the science of the overall programme were the 'Manuals For' handbooks, written in order to try to inject a degree of standardisation and comparability between studies, although researchers were never constrained to slavishly adopt these techniques, thus allowing their further development (Worthington 1975). Several of these IBP Manuals (e.g. Eleftheriou and McIntyre 2005) have continued to evolve into the present day, retaining their prime role in describing how to carry out research in particular systems.

An important feature of the IBP that resonates with the present emerging Ecosystem Approach is its inclusion of a social dimension – Human Adaptability. This is perhaps not too surprising given the focus of the programme – the production of food for a growing global human population. However, the linkages and feedbacks between social and ecological systems which characterise frameworks advocated today by, for example, the Millennium Ecosystem Assessment (MA) or the Convention on Biological Diversity (CBD), were not addressed. At the adoption of the Human Adaptability (HA) proposals in the early scoping meetings in the 1960s, 'a dissident view was voiced by the anthropologist Margaret Mead' (see Lutkehaus (2008) for a fascinating biography).

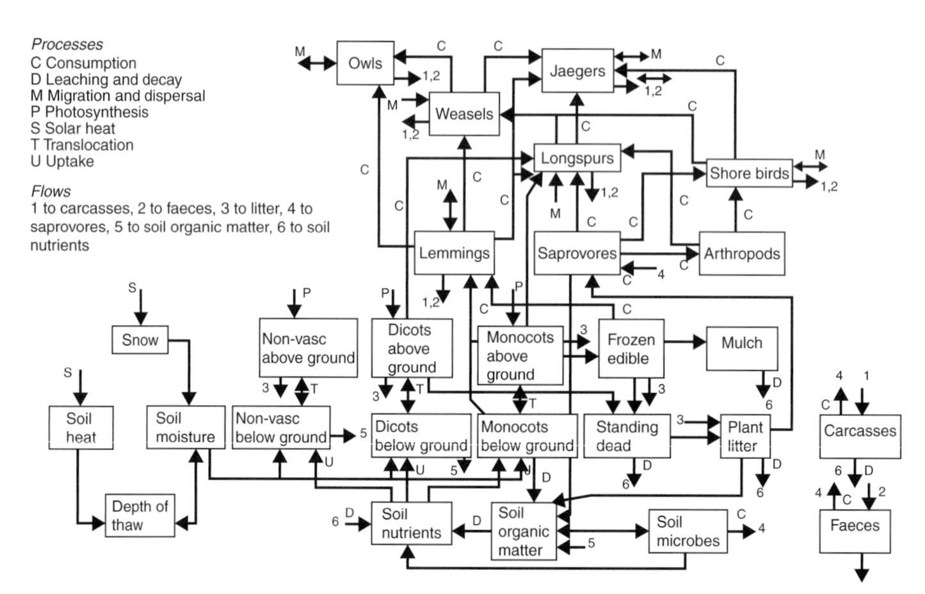

Figure 1.1 Box-and-flow diagram of a tundra ecosystem, Point Barrow, Alaska, typical of the representations used in IBP programmes to illustrate the relationships between key stocks of biomass. Adapted from Worthington (1975).

'The members listened with deep interest to Dr Mead's long and eloquent plea for the rejection of the HA proposals and the substitution of a programme based on the social sciences' (Collins and Weiner 1977, pp. 5–6). Whilst Mead's arguments are not recorded in detail, they were felt to be outside the scope of the programme and beyond the human biologists present, whose views ultimately prevailed. The Human Adaptation section of the IBP became concerned with surveys of the ability of humans to adapt to their environment in a social anthropology, physiological, genetic and medical sense, in an attempt to understand issues of health and welfare (e.g. growth and physique, genetic constitution, work capacity and pulmonary function, climatic tolerance, nutritional studies, medical and metabolic studies, demographic assessment) (Weiner and Lourie 1969). Whilst some way perhaps from the MEA and the Ecosystem Approach of the CBD, it should be remembered that the interdisciplinary approaches, paradigms and techniques we take for granted today were not as prominent, and in some cases did not even exist, in the 1970s and 1980s. The social dimension never seems to have achieved the emphasis it warranted within the IBP, perhaps because many areas of social science were not as fully developed as they are today or perhaps it was an idea whose time had not yet come.

Recently, the context and legacy of the IBP for current major international initiatives have been ably reviewed by Thomas Rosswall in his address to the British Ecological Society. Here, we restrict our analysis to the views expressed at the time by the US and the UK contributors in the context of what we might learn when designing future initiatives. The US efforts within the IBP dwarfed those of the UK in scale and funding. At its peak, 1,800 US scientists participated in the programme supported by $57 million in federal funds (Boffey 1976), an astonishing amount even by today's standards. Initially, it proved difficult to engage with all of the research community needed to deliver the programme, but, ironically, the programme suffered in the end from what one of the US planners described as 'ecological sprawl', as individual research studies only marginal to the original science vision signed up to be included under the IBP umbrella (Boffey 1968).

Other reported concerns were the lack of central governance of the science, within the US and for the programme as a whole (ibid.). Disappointingly, given the remit of the programme to examine the basis of productivity, agricultural research was largely ignored and, at least in the US, the Human Adaptability studies 'got relatively short shrift because they fell outside NSF's normal vision and the National Institutes of Health weren't interested' (Boffey 1976). Finally, although much was learned by the US ecosystem community as to how to work across the natural and physical sciences, one of the major science objectives, to develop systems-analysis models of ecosystems to assess human impacts and predict the effects of natural change 'largely failed, primarily because the goal was unrealistic in view of the lack of valid theory and experience in dealing with such large and complex systems' (ibid.).

A similar comment about systems analysis was made by Holdgate (in Worthington *et al.* 1976) in his assessment of the UK programme: the data demanded to construct systems-analysis models were underestimated and the ability to use those data was overestimated. The UK's assessment (see dedicated issue of *Philosophical Transactions of the Royal Society, series B*, volume 274 (1976)) pointed out other areas which could have developed better: there was too much compartmentalisation within studies and not enough cross-system comparison (Fogg and also Worthington, in Worthington *et al.* 1976); training of ecologists and knowledge exchange and transfer were not thought to have been achieved, especially in developing countries (Waddington and Worthington, in Worthington *et al.* 1976); there was no effective repository for the huge amounts of data collected (Worthington *et al.* 1976).

Reading the various IBP progress reports and post mortems, one is struck by the familiar and contemporary nature of many of the issues identified: the lack of overall programme governance; an unwillingness of some sections to become engaged at the start, and who therefore had little influence on the direction of the science; few plans for data storage and management and for final synthesis; a tendency of groups to work within those ecosystems with which they are most familiar and comfortable; issues of working across the disciplines, especially across the natural and social sciences. These all remain significant issues today for ecosystem ecology and the community needs to work hard to resolve them. Given the experience and lessons of the IBP, there can be no excuse for not anticipating such problems and putting mechanisms in place to deal with them.

Systems-analysis approaches

A major feature of the IBP, including the HA section described above, was the adoption of a systems-analysis approach. Thus, in their synthesis volume of the HA programme, Collins and Weiner (1977) state that 'The fruitfulness of this strategy – though it is costly in time resources and personnel – is well exemplified by the energy flow models developed in the American Andean project…the system serves to link calorie and nutrient exchanges with other population characteristics – the efficiency of work, the population density and the distribution of human biomass, etc.' A systems-analysis approach was thus recognised as demanding in resources (cf. appraisals by Boffey (1976) and Holdgate (1976), above), but it was deemed to have the capacity to link biological and social dimensions. Does this approach offer a way forward for prosecuting the Ecosystems Approach research agenda? To assess this we need to explore the context within which systems-analysis approaches to ecosystem questions have developed.

Many of the IBP programme synthesis volumes and related outputs contain a formal systems analysis, or at least a figurative representation of the major flows and components in a system using 'box-and-flow' diagrams, representing the biomass or state of a variable, and the flows representing inputs and outputs to and from other boxes (Figure 1.1). The degree to which such static representations help us to understand the dynamic nature of the system can be debated (remember, these were before the days of the personal computer or even the hand calculator), but they were helpful in representing the feedbacks and in identifying the major flows of material through the system.

At about the same time as the inception of the IBP, such holistic approaches were becoming familiar to a generation of ecologists through the extremely popular and influential *Fundamentals of Ecology* textbook by Eugene Odum (1953), and later with his brother Howard Odum (1959). H. T. Odum brought to the book his energy flow and thermodynamics approach, later formally presented as systems ecology in Odum (1983). Paul C. Stoy (this volume) provides an excellent account of this area. The brothers adopted a fundamentally holistic approach to their science that not only allowed an appreciation of the sources, sinks and flows of matter between ecosystem components, but also permitted an exploration of higher, ecosystem-level patterns and processes. Central to the school of thought that developed from, in particular, H. T. Odum's research group and associates is how these higher-level attributes change over time as the individual components, and hence the entire system, moves away from thermodynamic equilibrium through increased organisation and complexity of the components. Inevitably, much of the terminology and representation was borrowed from thermodynamic theory, including the notions of work, entropy and exergy. Systems analysis is thus a tool which allows identification of holistic properties of an ecosystem that can be achieved through a variety of applications. In the present ecosystem context, the most widely used are energy flow diagrams (e.g. Odum 1983) and various forms of ecological network analysis based on input–response–output theory (e.g. Patten *et al.* 1976, Ulanowicz 1986, 2000, Fath and Patten 1999).

Other terms and concepts needed to be developed as the science grew, such as ascendancy and energy (see also, Stoy, this volume). Ascendancy expresses the magnitude of the boxes-and-flows in the system (throughput) scaled by system complexity (information content), and has been shown to be a useful measure of ecosystem development state with links to stability (Christensen 1995). The concept of emergy (embodied energy) has been developed by H.T. Odum (Odum 1996, Odum and Odum 2000) and his colleagues (e.g. Costanza 1980) for addressing economic valuation aspects of environmental management and sustainability, so that energy can be represented in monetary terms.

The holistic approach, language and the use of heuristic devices and concepts such as ecosystem goals and directed development, inevitably set the systems school on a different trajectory from population biology, which is very much a reductionist science (e.g. Mansson and McGlade 1993). The tension between the reductionist and holistic camps has created considerable misunderstandings and misrepresentations, with an often bitter discourse. These different world views are reminiscent of the Clements–Gleason–Tansley debate, and are in part a reflection of different ways in which ecologists have historically approached their science in the UK and in North America. In a moving eulogy to H.T. Odum following his death in 2002, Brown *et al.* (2004) articulated very clearly the central issues. For those who had the privilege of working with Howard Odum, he was clearly an inspirational dynamo of a teacher. The price of being associated with this world view was their vilification and demonisation as 'Odumites' who promulgated 'Odumania', whilst some saw the holistic approach as somewhat 'blasphemous...and not to be trusted in a world where reductionism and small-scale biology held rein' (ibid.).

It is perhaps not surprising that the systems approach developed by Odum has been somewhat patchy in its geographical take-up. For instance, in a celebration of the oldest ecological society in the world, the British Ecological Society, and an assessment of the BES's contribution to the development of ecological ideas (Sheail 1987), Odum and his approach are not mentioned or referenced at all. This is by no means a criticism of John Sheail (his is a superb and comprehensive book), but a true reflection of how the relevance of this area has been perceived by what is a major and influential group of ecologists in the world. Two companion volumes produced by the BES for their jubilee celebration do contain three chapters: Waring (1989) on fluxes of matter and energy, Ulanowicz (1989) on thermodynamic-based approaches to oceans and Paul (1989) on soil processes (Cherrett 1989, Grubb and Whittaker 1989) but even today H.T. Odum's work and its legacy are not fully appreciated within the UK. The same is not true for other parts of Europe and for North America where, although there is the same reluctance by many ecologists to embrace this field if only as a heuristic device *sensu* Tansley (see above), the influence of Odum's ideas has been much more pervasive (but see also Stoy, this volume).

Whatever the issues, it is clear that the basic systems approach that Odum and others have advocated, and which the largest ecosystem programme to date, the IBP, embraced, has the potential for exploring the kind of dynamics and behaviour of large-scale systems which have recently come to the interest of policy makers. Many of these potentialities are encapsulated in Jorgensen *et al.* (2007), who have mapped applications of systems-based theory onto a broad variety of ecological areas including island biogeography, optimal foraging theory, niche theory, multipoint stability and diversity gradients. Jorgensen *et al.*

are motivated by the urgent need to develop a rigorous ecosystem theory given the expectations raised by the Millennium Ecosystem Assessment and the CBD's Ecosystem Approach. The book provides a primer for many areas of this approach to ecosystems which may be couched in terms unfamiliar to many readers, including an excellent account of Network Analysis. Importantly they include clarifications of many of the earlier criticisms of E.P. Odum's propositions and the unfortunate teleological-sounding terminology of ecosystem development, such as 'strategy', 'purpose' and 'goal functions'. It is made clear that these are seen as the consequences of internal feedback processes, not the drivers of such processes (ibid.).

Jorgensen *et al.*'s 'New Ecology' offers exciting and tantalising prospects for those in other areas of ecology to establish links and parallels between reductionist and holistic approaches, and has the potential to inform several areas within the Ecosystem Approach, as encapsulated by the Malawi Principles (Frid and Raffaelli, this volume). However, the New Ecology does not explicitly offer ways of fully coupling social and ecological systems, as would be demanded by a full implementation of the Ecosystem Approach.

Resilience thinking

Notwithstanding the difficulties some ecologists have with the driving forces behind and the correlates of ecosystem change in Odum's conjectures (Table 1.1), probably one of the more enduring concepts of ecosystems held by many researchers and policy makers is that (a) ecosystems develop over time through some kind of successional process, and (b) that development reaches an end-point system of some sort, as far from thermodynamic equilibrium as possible (see Stoy, this volume). Ecosystem management has traditionally been directed towards managing that developmental stage trajectory, all too often by attempting to maintain that end state around which our ideas of what is natural and desirable and which reflects many of our economic activities are focused.

An emerging view which seeks to challenge these preconceptions is Resilience Theory, of which there are two central themes: Environmental Thresholds and Adaptive Cycles (see Gunderson and Holling 2002 and Walker and Salt 2006 for excellent and accessible introductions to Resilence Theory). The reader may be more familiar with the concept of thresholds as alternate or multiple states or regime shifts (e.g. Steele 1996, Scheffer *et al.* 2001, Scheffer and Jeppesen 2007). Behind the resilience concept of thresholds is the involvement of 'slow' variables which change gradually over long time scales and which are thus difficult to detect. Examples would be the gradual accumulation of phosphorus in lake sediments and the slow rise in the salt-water table to the surface in converted landscapes found in Australia (Walker and Salt 2006). In both examples, all seems to be well with the system until the capacity of the system to absorb phosphorus is exceeded, or the salt water table reaches the roots of plants, at which point there are sudden and rapid changes in the system which has therefore crossed a threshold. Slow variables are difficult to track and detect because they have non-linear dynamics, they may be under the control of multiple factors and their slow progress means that managers experience the shifting baseline phenomenon (Pauly 1995), the curse of environmental management: often managers are only aware of the significance of the slow variable after a threshold has been crossed and large-scale ecological and social changes have occurred. Managing ecosystems to avoid such thresholds will therefore be challenging, but Groffman *et al.* (2006) offer some constructive suggestions with respect to critical pollution loads on natural systems.

Linked to the idea of thresholds is the concept of the Adaptive Cycle (Figure 1.2). The cycle has four phases: growth, conservation, release and reorganisation. The exploitation phase includes growth of the system to an end-state (conservation). Maintaining the system in that end-state often involves increasing optimisation and specialisation, which in turn increases the vulnerability of the system to external perturbations. The system inevitably collapses, the constituent capital is released and becomes available to be reorganised, possibly into a ecosystem similar to the original one, but perhaps along a different trajectory.

Adaptive cycles operate at all spatial and temporal scales, from individual leaves growing and dying on trees, to the life and death of entire stands of trees and forests (Figure 1.3). A key consideration for management is to ensure that cycles at different scales do not become synchronised over large areas, leading to the wholesale collapse of a resource like a forest, with its attendant massive ecological but also socio-economic change. The concepts within Resilience Theory seem to capture well the dynamics of a broad range of ecological, social and economic systems and have important implications for the management of coupled social–ecological systems. Ecological surprises (thresholds) are inevitable, but difficult to predict, and it makes little sense in the long run to attempt to maintain highly optimised systems at a desired and currently economically profitable end point. Doing so only increases the vulnerability of those systems and, therefore, of that social and economic activity.

A move towards more inclusive approaches to ecosystem ecology

Approaches to ecosystem ecology have involved a progressive engagement between the natural and physical sciences, as the need to investigate these systems within a loosely coupled biophysical framework has emerged. However, the majority of the world's ecosystems cannot be explored without explicit reference to the social systems that operate within, or impinge on, those biophysical entities. The main drivers of biodiversity change are people and the main recipients of healthy or unhealthy ecosystems are people: the two are intimately coupled. For example, between 1960 and 2000, the global population doubled to 6 billion people, as a result of which more land was converted

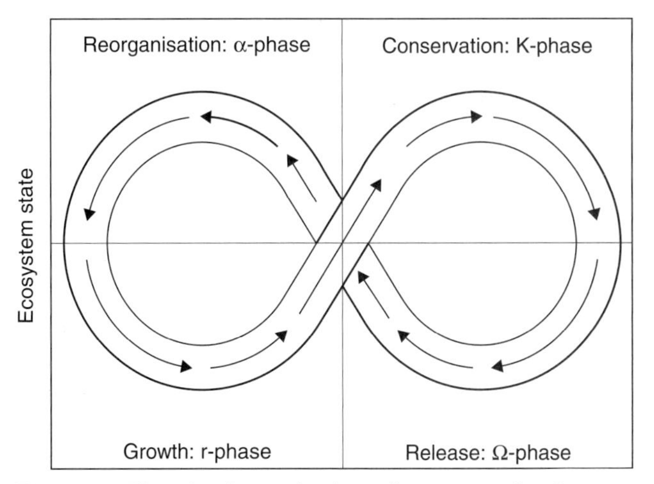

Figure 1.2 The adaptive cycle view of ecosystem development and change. In this perspective, collapse of the system is inevitable, whereupon the system components may re-assort and begin development again as a broadly similar system or one which is very different. Adapted from a number of sources.

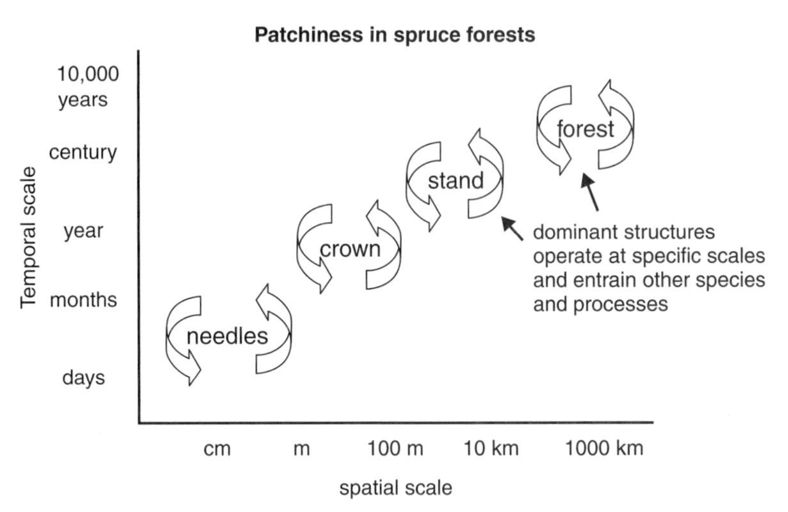

Figure 1.3 Patchiness occurs at a range of spatial and temporal scales in nature, as shown in this example of a spruce forest ecosystem. In such systems, dominant structures (from needles to forests) operate over different spatio-temporal scales. The cycles of life and death for each of these structures may follow adaptive cycle dynamics (see Figure 1.2), and these may entrain other ecological processes.

to agriculture between 1950 and 2000 than in the preceding 300 years (United Nations 2005). At a more regional scale, Piorr (2003) estimates that less than 3 per cent of the landscape of Europe remains un-dominated by agriculture, forestry or urban development. Similarly, it is unlikely that improving human welfare in line with the Millennium Development Goals can be achieved without acknowledging the co-dependence of people and the biophysical dimensions of their ecosystem (United Nations 2005).

There is a clear imperative, therefore, to understand what motivates people, whether they be individuals, communities, organisations or nation states, to manage ecosystems in the way they do, and to incorporate this knowledge into ecosystem models in order to explain ecosystem structure and change. This is the present phase of the evolution of ecosystem ecology and one which will require a far greater relaxing of the term 'ecosystem' in order to accommodate those additional disciplines whose business it is to understand human motivation, such as economists, sociologists, anthropologists and psychologists. Not all mainstream ecologists are comfortable with this, although we hope the chapters within this volume will help to convince them of the need to work with and across other disciplines. Nevertheless, interdisciplinary working within holistic approaches to ecosystem research and management does bring with it a number of challenges. Jones and Paramor (this volume) discuss the wider questions such working throws up for those wishing to travel this road, not least the challenges it presents to engrained ways of thinking and value systems held by those coming from different backgrounds. In addition, there are real, operational issues that need to be overcome if ecosystem research and management programmes are to be successful.

A fundamental question is what added value does the interdisciplinary approach deliver that single-discipline or many-discipline approaches cannot? This seems an entirely reasonable question: why should such an approach be superior to, for instance, biophysical researchers and social science researchers working independently and then presenting their findings to a third party who is then responsible for making a decision about how best to manage that ecosystem? The answer to that question turns out to be surprisingly difficult to formulate rigorously. To our knowledge, there has been no 'experiment' where programmes have been embarked upon from both the interdisciplinary and mono- or multidisciplinary perspectives and recommendations compared. Nevertheless, we think there is compelling evidence and argument for the superiority of an interdisciplinary approach. The National Academy of Sciences (2004) carried out a thorough review of programmes that had been set up to tackle the most difficult issues of the day including the development of the atomic bomb and putting a man on the moon. It is clear from their report (which includes many other examples) that it would have been extremely difficult to arrive at solutions to those pressing problems without an interdisciplinary approach, in particular a mutual understanding of and by the disciplines required. We can think of no more pressing problem for the planet at present than the need to develop good practice in sustainability based on fundamental research and knowledge.

In addition to the historical successes of interdisciplinary programmes, a case can be made that other approaches based largely on single-discipline or at best multidisciplinary approaches, simply haven't worked: ecosystem deterioration continues apace (United Nations 2005). Only time will tell if interdisciplinary approaches will be more successful, but they could hardly fare worse and at least they explicitly recognise the interdisciplinary nature of the close coupling of social and biophysical processes.

Jones and Paramor (this volume) review the career development and cultural barriers that need to be overcome if researchers are to be encouraged to adopt an interdisciplinary approach and we will not rehearse these again here. Instead, we briefly reflect on what would make interdisciplinarity work within the next generation of ecosystem programmes. Some pointers are available from forward-looking national and international programmes established several years ago, such as DIVERSITAS, ICRAF (International Centre for Research in Agroforestry), WorldFish Centre (formally ICLARM) and IHDP (International Human Dimensions Programme on Global Environmental Change).

A review of the mechanisms that best facilitated and built interdisciplinary capacity within such programmes revealed a number of consistent features (Raffaelli 2006, White *et al.* 2009). Time and financial resources were, of course, important, but way ahead of these were effective leadership, a mutual respect for each other's disciplines and an unswerving commitment of participants. Interdisciplinary programmes on ecosystem research and management will need to ensure that these criteria are met if they are to be successful.

Interestingly, many respondents declared similar barriers to interdisciplinary research as those identified by Jones and Paramor (this volume): inappropriate career rewards in academia for interdisciplinary research; a culture of competition between mono- and interdisciplinary funding from the funding councils and within academic institutions; inappropriate orientation of research evaluation procedures towards monodisciplinary research; poor interpersonal/interdisciplinary relationships and lack of trust; the greater time and energy required for interdisciplinary compared to monodisciplinary research. Specific mechanisms for facilitating interdisciplinarity included: the use of clear language without jargon; regular face-to-face informal meetings; availability of forums to facilitate discussions, regular self-evaluation and learning sessions; and the use of participatory methods. These help to establish a team philosophy, promoted further by sharing of all information and data freely and agreeing ground rules on intellectual property in advance, especially authorship of published outputs. All of these are ways to build commitment, trust and respect between researchers from different disciplines, attributes which together with leadership, appear to contribute greatly to the success of programmes.

Table 1.1. *Mean scores (/6) of 44 respondents' experience of how aspects of interdisciplinary research influenced interdisciplinary success within their international programme or project (scoring: 6-strongly represented, 1-not represented). After Raffaelli (2006) and White* et al. *(2009).*

Rank	Influence of factor	Mean score
1	Leadership	5.36
2	Commitment	5.14
3	Common frameworks	4.93
4	Respect	4.82
5	Trust	4.68
6	Relationships	4.58
7	Time	4.39
8	Funds	4.28
9	Common learning	4.27
10	Inclusiveness	4.05
11	Negotiation of roles	3.64
12	Teambuilding	3.45
13	Close proximity	2.64

What can be learned from the history of ecosystem ecology?

It is clear that ecosystem ecology as a concept and a discipline has travelled a long way since Tansley's day, becoming broader and more inclusive in its scope, although the landscape over which it has passed has been decidedly rocky in places. New approaches and world views have emerged to challenge conventional wisdom and these have inevitably challenged prevailing paradigms and often generated a hostile audience. The increasing emphasis on the interdependence of natural and social systems demands holistic approaches involving researchers from many different disciplines, including mainstream ecology, which has a huge amount to offer. The environmental problems which need to be addressed are often driven by processes operating at regional and global scales, so that large-scale programmes with a matching large-scale vision will be required to address those problems. We should learn from the lessons of previous large-scale, visionary initiatives programmes, such as the International Biological Programme. Society is unlikely to forgive us for ignoring those lessons.

Acknowledgements

The authors are indebted to valuable comments, suggestions and insights from Bob Ulanowicz, Alastair Fitter and Thomas Rosswall.

References

Boffey, PM (1968). International Biological Programme: U.S. effort stands on shaky ground. *Science* **159**: 1331–4.

Boffey, PM (1976). International Biological Programme: was it worth the cost and effort? *Science* **193**: 866–8.

Brown, MT, Hall, CAS and Jorgensen, SE (2004). Eulogy. *Ecological Modelling* **178**: 1–10.

Cherrett, JM (1989). *Ecological Concepts: the contribution of ecology to an understanding of the natural world*. Blackwell Scientific Publications, Oxford.

Christensen, V (1995). Ecosystem maturity – towards quantification. *Ecological Modelling* **77**: 3–32.

Collins, KJ and Weiner, JS (1977). *Human Adaptability. A history and compendium of research*. Taylor & Francis Ltd, London, St Martin's Press Inc., New York.

Convention on Biological Diversity (2009). http:/www.cbd.int/ecosystem/ (sourced January 2009).

Costanza, R (1980). Embodied energy and economic valuation. *Science* **210**: 1219–24.

Eleftheriou, A. and McIntyre, AD (2005). *Methods for the Study of Marine Benthos*. Blackwell Publishing Limited, Oxford.

Elton, C (1927). *Animal Ecology*. Sidgwick & Jackson, London.

Fath, BD and Patten, BC (1999). Review of the foundations of network environ analysis. *Ecosystems* **2**: 167–79.

Groffman, PM, Baron, JS, Blett, T, Gold, AJ, Goodman, I, Gunderson, JH, Levinson, BM, Palmer, MA, Paerl, HW, Peterson, GD, Poff, NL, Rejeski, DW, Reynolds, JF, Turner, MG, Weathers, KC and Weins J (2006). Ecological thresholds: the key to successful environmental management or an important concept with no practical application? *Ecosystems* **9**: 1–13.

Grubb, PJ and Whittaker, JB (1989). *Toward a More Exact Ecology*. Blackwell Scientific Publications, Oxford.

Gunderson, LH and Holling, CS (2002). *Panarchy: understanding transformations in human and natural systems*. Island Press, Washington DC.

Jax, K (2007). Can we define Ecosystems? On the confusion between definition and description of ecological concepts. *Acta Biotheoretica* **55**, 4: 341–55.

Jorgensen, SE, Fath, BD, Bastianoni, S, Marques, JC, Muller, F, Nielson, SN, Patten, BC, Tiezzi, E and Ulanowicz, RE (2007). *A New Ecology. Systems Perspective*. Elsevier, Amsterdam.

Lindeman, RL (1942). The trophic-dynamic concept of ecology. *Ecology* **23**: 399–418.

Lutkehaus, NC (2008). *Margaret Mead: the making of an American icon*. Princeton University Press.

Mansson, BA and McGlade, JM (1993). Ecology thermodynamics and Odum's conjectures. *Oecologia* **93**: 582–96.

National Academy of Sciences (2004). *Facilitating Interdisciplinary Research*. National Academy of Sciences.

Odum, EP (1953). *Fundamentals of Ecology*, 1st edn. W.B. Saunders, Philadelphia.

Odum, EP and Odum, HT (1959). *Fundamentals of Ecology*, 2nd edn. W.B. Saunders, Philadelphia.

Odum, HT (1983). *Systems ecology*. Wiley, New York.

Odum, HT (1996). *Environmental Accounting. EMERGY and environmental decision making*. John Wiley & Sons, New York.

Odum, HT and Odum, EP (2000). The energetic basis for valuation of ecosystem services. *Ecosystems* **3**: 21–3.

Patten, BC, Bosserman, RW, Finn, JT and Cale, WG (1976). Propagation of cause in ecosystems. In Patten, BC (ed.), *Systems Analysis and Simulation in Ecology*, Vol. 4. Academic Press, New York, pp. 457–9.

Paul, EA (1989). Soils as components and controllers of ecosystem processes. In

Grubb, PJ and Whittaker, JB (eds.). *Toward a More Exact Ecology*. Blackwell Scientific Publications, Oxford, pp. 327–52.

Pauly, D (1995). Anecdotes and the shifting baseline syndrome of fisheries. *Trends in Ecology and Evolution* **10**(10): 430.

Piorr, H-P (2003). Environmental policy, agri-environmental indicators and landscape indicators. *Agriculture, Ecosystems & Environment* **98** (1–3): 17–33.

Raffaelli, D (2006). RELU: The International Context. Final report. http://www.relu.ac.uk/research/projects/EOASummaries/RaffaelliEOASummary.pdf

Scheffer, M, Carpenter, SR, Foley, JA, Folke, C and Walker, B (2001). Catastrophic shifts in ecosystems. *Nature* **413**: 591–6.

Scheffer, M and Jeppesen, E (2007). Regime shifts in shallow lakes. *Ecosystems* **10**: 1–3.

Sheail, J (1987). *Seventy-Five Years in Ecology: The British Ecological Society*. Blackwell Scientific Publications, Oxford.

Steele, JH (1996). Regime shifts in fisheries management. *Fisheries Research* **25**: 19–23.

Tansley, AG (1914). Presidential Address. *Journal of Ecology* **2**: 194–203.

Tansley, AG (1935). The use and abuse of vegetational concepts and terms. *Ecology* **16**: 284–307.

Ulanowicz, RE (1986). *Growth and development: ecosystems phenomenology*. Springer Verlag, New York, p. 203.

Ulanowicz, RE (1989). Energy flow and productivity in the oceans. In Grubb, PJ and Whittaker, JB (eds.). *Toward a More Exact Ecology*. Blackwell Scientific Publications, Oxford, pp. 327–52.

Ulanowicz, RE (1997). *Ecology, the Ascendent Perspective*. Columbia University Press, New York.

Ulanowicz, RE (2000). Ascendancy: a measure of ecosystem performance. In Jørgensen, SE and Miller, F (eds.). *Handbook of ecosystem theories and management*. CRC Press, Boca Raton, Florida, pp. 303–16.

United Nations (1992). *Convention on Biological Diversity*. UN, New York.

United Nations (2005). *Millennium Assessment. Living beyond our means: natural assets and human well-being*. UN, New York.

Walker, B and Salt, D (2006). *Resilience Thinking. Sustaining ecosystems and people in a changing world*. Island Press, Washington DC.

Waring, RH (1989). Ecosystems: fluxes of matter and energy. In *Ecological Concepts: the contribution of ecology to an understanding of the natural world*. Blackwell Scientific Publications, Oxford, pp. 17–41.

Weiner, JS and Lourie, JA (1969). *Human Biology: a guide to field methods*. Blackwell Scientific Publications, Oxford.

White, PCL, Cinderby, S, Raffaelli, D, de Bruin, A, Holt, A and Huby, M (2009). Enhancing the effectiveness of policy-relevant integrative research in rural areas. *Area* **41**: 414–424.

Willis, A J (1997). The ecosystem: an evolving concept viewed historically. *Functional Ecology* **11**: 268–71.

Worthington, EB (1965). The International Biological Programme. *Nature*, **208**, 223–6.

Worthington, EB (1975). *The Evolution of the IBP*. Cambridge University Press.

Worthington, EB, Fogg, GE, Waddington, CH, Clymo, RS, Newbould, PJ, Holdgate, MW and Clarke, C (1976). General Discussion. *Philosophical Transactions of the Royal Society of London, Series B* **274**: 499–507.

Worthington, EB (1983). *The Ecological Century. A Personal Appraisal*. Clarendon Press, Oxford.

CHAPTER TWO

Linking population, community and ecosystem ecology within mainstream ecology

ANDY FENTON and MATTHEW SPENCER
School of Environmental Sciences, University of Liverpool

Introduction

Charles Darwin's tangled bank provides one of the best-known early descriptions of an ecosystem:

> It is interesting to contemplate a tangled bank, clothed with many plants of many kinds, with birds singing on the bushes, with various insects flitting about, and with worms crawling through the damp earth, and to reflect that these elaborately constructed forms, so different from each other, and dependent on each other in so complex a manner...

This description highlights much of the complexity of ecosystems, comprising various biotic components (plants, vertebrates and invertebrates), abiotic factors (soil) and environmental conditions (humidity). Even though this list comprises only a fraction of the likely diversity within the ecosystem, and Darwin has combined many individual species into single groups (plants, birds, insects), the stated inter-dependencies emphasise the large number of potential direct and indirect interactions that may occur among the various components, and between them and the environment.

Understanding the functioning of ecosystems, determining the factors underlying their structure and predicting their responses to perturbations are major challenges that have formed the lifeblood of population, community and ecosystem ecology for decades. Given the daunting task of addressing these issues empirically, mathematical models have played a vital role in this research. By simplifying the complexity of an ecosystem, formalising hypotheses and often producing testable predictions, mathematical models can provide invaluable insights into the processes shaping ecosystems and, from an applied perspective, inform the development of management policies (e.g. species conservation, land management, harvesting regimes). The challenges facing us in the twenty-first century, including increased rates of biodiversity loss, global climate change and the potential emergence of novel infectious diseases, emphasise that the need to develop a deep understanding of ecosystem ecology has never been greater. It is here that mathematical models could play a vital role.

Ecosystem Ecology: A New Synthesis, eds. David G. Raffaelli and Christopher L. J. Frid. Published by Cambridge University Press.

However, population, community and ecosystem ecologists have all developed their own ways of exploring questions at their scale of interest (Raffaelli and Frid, this volume). Although each approach may be informative within a specific level, it is not clear how they scale between levels, whether there are discontinuities or whether there are commonalities at different scales that suggest the potential for synergistic interaction between these fields of research. Here we provide an overview of theoretical approaches developed at each scale, highlighting their similarities and differences, before describing various approaches that may provide some degree of connection between them. A key theme is that statistical evaluations and simplifications will be essential if population-level models are to be useful in understanding and, ultimately, predicting the behaviour of ecosystems.

An overview of population-, community- and ecosystem-level modelling approaches

The theoretical approaches at the three scales differ in a number of respects, ranging from their complexity, their level of abstraction, whether they are dynamic or static, whether they are intended to be heuristic or truly predictive and the extent to which they are dependent on ecological data. These differences arise primarily from the types of question they are constructed to address. In this section we describe the main characteristics of existing modelling approaches at each of these scales. A technical description of a generalised modelling framework that is used to provide the basis of many of the approaches discussed later in this chapter is given in the box.

Box 2.1 A generalised model

The most general model we will consider is a continuous-time, stochastic state-space model for the dynamics of each of $i=1 \ldots m$ species. We use continuous-time examples for the most part, but similar ideas apply to discrete-time models.

The dynamics of species i can be described by a stochastic differential equation:

$$dx_i = \left[f_i(x_i) + \sum_{j=1}^{m} g_{ij}(x_i x_j) \right] dt + h_i(x_i) dW_i(t) \tag{1}$$

where x_i is the abundance or density of species i, $f_i(.)$ is an unspecified function giving the growth rate of species i and $g_{ij}(.)$ describes the effect of species j on the growth rate of species i. The term $h_i(x_i)dW_i(t)$ is a noise term where $W_i(t)$ is a random variable whose value at time t is determined by a standard Wiener (Brownian motion) process, in which increments $W_i(t) - W_i(s)$ (where $s < t$) are normally distributed with zero mean and variance $t - s$ (Higham 2001). The unspecified function $h_i(.)$ determines how noise affects the dynamics of species i, and is assumed to represent the effects of all state variables not explicitly modelled.

Equation 1 is known as a *state equation*. However, the true state of the system may not be directly observable and so, in addition, we have a *measurement equation* that represents what we actually observe. We assume that the system is only observed at discrete time points, and so the observed abundance (or density, or biomass etc.) of the ith species at time t, $y_i(t)$, is:

$$y_i(t)=\phi_i(x_i(t)) \tag{2}$$

where $\phi_i(.)$ is an unspecified function describing how the observed density of species i is related to its true density. Together, Equations 1 and 2 define a *state-space model* for the system (Harvey 1989, section 3.1).

We define the vector $\mathbf{x}_i^{\mathrm{T}}=\{x_i(0)\ldots x_i(\mathbf{T})\}$ as the true densities of species i at all observation times from 0 to T, and $\mathbf{y}_i^{\mathrm{T}}=\{y_i(0)\ldots y_i(\mathbf{T})\}$ as the observed densities of species i from times 0 to T. Furthermore, we will write $\mathbf{X}^{\mathrm{T}}=\{\mathbf{x}_1^{\mathrm{T}}\ldots\mathbf{x}_m^{\mathrm{T}}\}$ for the set of true densities of all species at all observation times, and $\mathbf{Y}^{\mathrm{T}}=\{\mathbf{y}_1^{\mathrm{T}}\ldots\mathbf{y}_m^{\mathrm{T}}\}$ for the set of all observed densities of all species at all times. Finally, this model has a vector of p parameters, $\boldsymbol{\theta}=\{\theta_1\ldots\theta_p\}$, incorporating initial densities and coefficients used in the unspecified functions.

Population-level models

The use of mathematical models in the analysis of population dynamics dates back at least a century and represents a prime example of how mathematics can be used to shed light on natural phenomena. Population models typically describe the dynamics of relatively few species over time and so tend to be of low dimensionality, with relatively few state variables and parameters. Furthermore, these models tend to be deterministic (i.e., $h_i(.)=0$ in Equation 1), ignoring environmental and demographic stochasticity and so the predicted dynamics emerge from underlying demographic processes of an 'average' individual in the population. In addition they are frequently non-linear (e.g. incorporating density dependencies) and potentially lead to complex dynamics. Finally, these models are primarily 'biotic', describing species dynamics while ignoring various abiotic and environmental variables. Overall, such models are typically highly generic, and often are primarily intended to aid understanding rather than being truly predictive.

Classic examples of differential equation population-level models are the Lotka–Volterra predator–prey and competition models, originally developed in the early twentieth century. These models are obtained from Equation 1 by setting $f_i(.)=a_i x_i$, $g_{ij}(.)=\alpha_{ij} x_i x_j$ and $h_i(.)=0$. Hence, the general model for a community of m species may be written as:

$$\frac{dx_i}{dt}=x_i\left(a_i+\sum_{j=1}^{m}\alpha_{ij}x_j\right) \tag{3}$$

where x_i is the abundance (or density, or biomass) of species i, a_i is the *per capita*, density-independent rate of increase (or decrease) of species i and α_{ij} is the *per capita* strength of interspecific interaction of species j on species i. These models are typically analysed using standard analytical methods that tell us about general properties of the system, including the presence of equilibrial states, their stability (i.e. whether they exhibit stable point equilibria, sustained limit cycles or unstable, divergent oscillations) (e.g. May 1974, chapter 2). As such, they are very useful in allowing exploration of how various processes (e.g. density-dependent growth, non-linear functional responses, time lags etc.) alter the stability and dynamics of natural populations and communities.

Community-level models

The distinction between population-level models and community-level models is rather blurred; models that adopt the population-level approach described above can be used to describe the interactions between relatively large numbers of species. One distinction may be that population-level models are typically based on the aggregative behaviour of individuals (i.e. comprising individual birth and survival rates), whereas community-level models are based on the broader properties of the populations. Hence, while population dynamic models tend to adopt relatively mechanistic approaches to explore species dynamics, community ecology tends to adopt a more phenomenological approach to determine the factors leading to observed patterns in community structure (e.g. community diversity indices, species abundance distributions, biomass pyramids or network topology). These observed patterns can be compared to patterns generated from models of artificial communities, making it possible to determine whether the biological mechanisms built into the models are capable of generating realistic-looking communities. Hence the distinction between population-level and community-level approaches has more to do with the types of question being asked, rather than simply the number of species being considered.

A classic example of the community-level approach concerns the relationship between community complexity and stability, and this ongoing debate has recently been rekindled due to concerns about rates of biodiversity loss and global climate change. Clearly, community stability is a crucial factor underlying biodiversity, and understanding the factors determining the stability of a given community is essential for preventing species declines following the loss of other species (Ebenman and Jonsson 2005). Intuition suggests that more 'complex' communities (possibly those with more species or more links between species) should be more stable than simple communities, since they can buffer the community from perturbations (MacArthur 1955, Elton 1958). For example, predators with many prey species should be less affected by loss of one of those species than predators with very few prey species. However, this simple view has been repeatedly challenged since it was first proposed over fifty years ago, with different relationships between community complexity and stability being found depending upon how complexity and stability were defined (Pimm 1984). For example, early models of randomly-assembled communities showed that more complex communities (those with more species, greater connectance, or greater mean interaction strength) are, at best, no more inherently stable than simpler communities (May 1974, chapter 3). However, more recent community models have shown that stability is, to a large extent, not determined by community complexity *per se*, but by not only the strength, but also the distribution of links throughout the community (Jansen and Kokkoris 2003). These studies suggest that community stability is enhanced by the presence of many weak links that bind the community together (McCann *et al.* 1998, Proulx *et al.* 2005). Furthermore, the distribution of interaction strengths can have important consequences for ecosystem functioning (Duffy 2002, Montoya *et al.* 2003). Such theoretical predictions are finding support from empirical studies (de Ruiter *et al.* 1995, Neutel *et al.* 2002), suggesting that natural communities are structured in a specific way that confers stability; presumably communities not structured in this way rapidly disintegrate and are rarely observed.

Ecosystem-level models

We define an ecosystem-level model as one which incorporates the interrelationships between both biotic (e.g. species) and abiotic factors (e.g. nutrients and other components that affect ecosystem functioning) within a single framework. Hence, in one sense they may be considered as population- or community-level models in which some of the state variables of Equation 1 represent abiotic components. Therefore, ecosystem-level models tend to be concerned with the flow of energy among the various biotic and abiotic compartments of the ecosystem (DeAngelis 1992, Loreau 2000). Such models may also be embedded into wider models defining the ecosystem's physical environment (e.g. describing changes in tidal or sea circulation patterns for marine ecosystem models; Neumann 2000). As such, ecosystem models are typically highly detailed, of high dimension, with large numbers of state variables and parameters, and often are tailored towards specific ecosystems with the intention of addressing specific, applied questions.

The use of ecosystem models has increased greatly over the last few decades due to ever-increasing computational power, allowing models to be increasingly detailed. This leads to a fine balancing act between the benefits, in terms of making very detailed specific predictions, of developing a highly complex model and the costs, in terms of making the models unwieldy and difficult to interpret. In particular, model parameterisation of complex models can be a daunting task, resulting in very detailed models being developed that are ultimately based on less detailed biological knowledge. Furthermore, interpreting their predictions and evaluating their outputs can be challenging – the large numbers of parameters in these models and the potentially long chains of indirect interactions can make it difficult to determine the mechanisms underlying any observed response. Finally, there is a danger of unquestioningly believing the models due to their complexity; the underlying assumptions of simple models are frequently questioned due to their obvious limitations. However, the limiting assumptions of more complex models are often far less apparent and it is easy to forget that even the most detailed models are still simplifications of reality, potentially ignoring many key biological processes. Despite these caveats though, ecosystem models provide an invaluable framework for studying the dynamics of ecosystems, and their use is only likely to increase in the coming years. They are capable of being both heuristic tools, providing insight into the key properties of generalised ecosystems, and of being of genuine applied use, able to make specific predictions that can shape ecosystem management decisions in a way that more simple population or community models cannot (Fulton *et al.* 2003).

Possibly the most common application of ecosystem models is to the study of marine ecosystems, often with the intention of understanding how global climate change might affect their functioning. This has led to the emergence of a variety of modelling approaches and an impressive array of acronyms: names such as ECOPATH, ECOSIM, IGBEM and so on, are commonplace within many marine ecology journals (Christensen and Pauly 1992, Heymans and Baird 2000, Pauly *et al.* 2000, Christensen and Walters 2004, Fulton 2004, Petihakis *et al.* 2007, Haputhantri *et al.* 2008). One of the most widely adopted ecosystem models is ERSEM (European Regional Seas Ecosystem Model; Baretta *et al.* 1995, Barettabekker *et al.* 1995, Blackford *et al.* 2004). ERSEM is a generic model that simulates the temporal and spatial dynamics of the 'microbial loop', comprising bacteria, phyto- and zooplankton, together with the flow of dissolved and particulate organic matter and essential nutrients such as carbon, nitrogen and silicate in the ocean (Blackford *et al.* 2004). These components are typically placed into 'functional', rather than taxonomic groups, based on cell size (e.g. picoflagellates, microzooplankton, mesozooplankton) in which the key biological properties of organisms within a group are approximately the same. In this way the model simplifies a highly complex ecosystem, while retaining many of the key processes that may have a significant impact on ecosystem dynamics.

The ERSEM model may be applied to a range of scenarios, including coastal, oligotrophic or eutrophic situations, incorporating various spatial heterogeneities and incorporating a number of key environmental drivers, such as temperature, irradiance, cloud conditions and mixing due to wind. As such the model may be used either heuristically or predictively, when tailored to specific ecosystems. Indeed, this flexibility is one of its key features. The biochemical component of the model utilises a standardised set of parameters and biological processes that are often kept constant across a range of scenarios (Blackford *et al.* 2004). However, this standard model may then be placed within a specific physical model describing tidal flows, spatial heterogeneities and climatic conditions, thereby tailoring the model to the circumstance of interest. The model was originally developed for the North Sea, but has since been applied to other habitats, including estuarine, temperate (the Mediterranean Sea) and monsoonally forced tropical environments (the Arabian Sea) (Allen *et al.* 1998, Vichi *et al.* 1998, Allen *et al.* 2002, Blackford and Burkill 2002). However, as we discuss below, determining how well such complex models perform is far from a trivial task.

Model evaluation

A central theme in this chapter is that in order to build workable models of ecosystems, we need to simplify a potentially very complex ecological system. However, in order to decide which simplifications are appropriate, we need a way of evaluating each one (i.e. determining whether a given model is a good description of a set of observed data). In this section, we review what ecologists do with models, and the appropriate ways to evaluate models of the form given by Equations 1 and 2. We show that what ecologists do with models is systematically different from what molecular phylogeneticists do with models. We argue that quantitative evaluation of the fit of models to data is central to the process of model improvement, to the extent that without it, we are admitting that our models are not supposed to be any use.

How do ecologists evaluate models?

Scientists are fond of diagrams showing a cyclical process of theory development, data collection, theory evaluation and modification. Despite this, some areas of population biology have developed in a more-or-less data-free way, and this may have held back the development of ecological models. To illustrate this, we reviewed papers published in 2005 in the journal *Ecology*, one of the leading general ecology journals. We used ISI Web of Science to find all papers containing the keyword 'model'. We then categorised each paper by the way it evaluated models and the sources of data for this evaluation. For our purposes, we defined a model as a quantitative description of a biological process, specifying how the process generates data. For comparison, we used the same methods to review papers published in 2005 in the journal *Molecular Biology and Evolution.*

Table 2.1. *Data sources and evaluation of models in the journals* Ecology *and* Molecular Biology and Evolution.

	Ecology	*Molecular Biology and Evolution*
Contained models[1]	66	52
Contained no data	12 (18%)	8 (15%)
Explicit reference to database use	2 (3%)	28 (54%)
Collected new empirical data	31 (47%)	18 (35%)
Evaluated models:		
not at all	13 (20%)	7 (13%)
qualitatively	20 (30%)	2 (4%)

[1] Out of papers published in 2005 recovered using ISI Web of Science with keyword 'model', and judged to contain models according to the criteria described in the text.

We chose *Molecular Biology and Evolution* because it is one of the key journals in the field of molecular phylogenetics. There has been spectacular progress in the last thirty years in building complex models of evolutionary processes, and using these models to reconstruct evolutionary trees. We then consider how these two fields compare in terms of how they combine models and data.

We found 66 papers in *Ecology* and 52 in *Molecular Biology and Evolution* that met our search criteria and contained models according to our definition (Table 2.1). We used chi-square tests to compare their distributions across several categorical variables. Out of these papers, the proportions containing no data are similar between the two journals (χ^2=0.02, df=1, P=0.81). Papers in *Ecology* are less likely to make explicit reference to database use (χ^2=36.98, df=1, P=1.20e^{-9}), but the proportions reporting new empirical data are similar between the two journals (χ^2=1.35, df=1, P=0.24). The distributions of ways in which models are evaluated are very different between the two journals (χ^2=16.41, df=2, P=0.0003), largely because papers in *Ecology* rely on qualitative evaluation more often than expected, and papers in *Molecular Biology and Evolution* less often than expected.

We hesitate to generalise too much from this sample of two journals in a single year. However, these preliminary results do suggest some possible differences in the culture of model use. Perhaps one reason that ecologists are reluctant to formally evaluate their models is the perception that ecosystems are simply too complex to be realistically modelled. Molecular biologists apparently felt the same way about early models of molecular evolution, forcing the pioneering modellers to hide their work within empirical papers (Felsenstein 2001). However, perhaps ecologists can borrow from current molecular biology approaches to model development and subsequent evaluation. In particular, although parameterising a model that describes the average rate of substitution of one amino acid for another is a complicated task (Whelan and Goldman 2001), it need only be done once, and the results can be applied to a huge number of different protein sequences. Similarly, in ecology it may be possible to use average parameter values, or values derived from other, related species, or values derived from known allometric relationships (e.g. growth, reproduction, mortality or feeding rates based on body size) to make accurate predictions of specific ecosystems. This is precisely what the ERSEM model does; by placing a common biochemical model with a standard set of parameter values into a tailored physical model it is possible to make predictions about specific ecosystems. However, once such a model has been constructed it becomes essential to rigorously evaluate it to determine how well it fits a set of observations.

Statistical evaluation of models

Many models in population biology are derived from Equations 1 and 2 by setting $h_i(.)=0$ (so that they are deterministic) and $\phi_i(.)=1$ (so that there is no observation error). In such cases, any differences between the model and observed data can only be interpreted as inadequacies in model structure. This does not help us to simplify models. The sum of squared deviations between the output from a deterministic model and a dataset is a popular way of evaluating a model (e.g. Harrison 1995). However, if we have not specified the functions $h_i(.)$ and $\phi_i(.)$, we do not know whether this is the right criterion. Furthermore, we have no theory that tells us whether the addition of a new parameter significantly improves the fit of the model. We may imagine alternative criteria that also seem important, such as the match of the predictions to the autocorrelation function or spectral density of an observed time series. We might try to combine many such criteria (Kendall *et al.* 1999), but without theory, we have no way of knowing how to weight each one.

Fortunately, there has been substantial progress in fitting state-space models to data. Given a model $\boldsymbol{M}$ that specifies both environmental stochasticity and measurement error, a set of data $\mathbf{Y}^T$ and a parameter vector $\boldsymbol{\theta}$, we can use computationally intensive methods such as particle filters to estimate the likelihood $L(\boldsymbol{\theta};\mathbf{Y}^T, \boldsymbol{M})$ by integrating over the unknown true states of the system (Buckland *et al.* 2004). If we estimate $\boldsymbol{\theta}$ from the data, we can in principle use likelihood ratio tests (Bickel and Doksum 2001, section 6.3.1) or Akaike's Information Criterion (AIC) (Bozdogan 1987) to make informed choices about model simplification. In essence, we compare the likelihoods $L(\boldsymbol{\theta}_0;\mathbf{Y}^T, \boldsymbol{M}_0)$ and $L(\boldsymbol{\theta}_1;\mathbf{Y}^T, \boldsymbol{M}_1)$ for two models $\boldsymbol{M}_0$ and $\boldsymbol{M}_1$ with associated parameter vectors $\boldsymbol{\theta}_0$ and $\boldsymbol{\theta}_1$, and use statistical theory to decide whether the improvement in fit from the more complex model is large enough to justify the extra parameters it uses. In practice this is complicated because for many complex models we cannot obtain exact likelihoods. Nevertheless, framing the discussion in these terms is useful because it makes clear what is being compared. Therefore ecologists need to embrace these methods of combining theory and data if they are to complete the model development–evaluation–modification cycle and make truly informative and useful models.

Suggested approaches for scaling between population and ecosystem ecology

Clearly ecosystems are highly complex, and producing accurate models that capture all the subtle interactions occurring within them is unlikely to be feasible, or even desirable, for all but the simplest of ecosystems. However, simple population-level models are unlikely to have sufficient detail to make specific predictions about natural systems. Therefore questions remain as to whether it is possible to scale up from simple population-level models or scale down further from complex ecosystem models and whether there is some intermediate level of complexity that allows genuine predictions to be made without sacrificing accuracy. Here we describe four methods that have been suggested for simplifying ecosystem models, which may allow such issues to be addressed.

Linearised, equilibrium approaches

One means of bringing population-level modelling approaches into community and ecosystem ecology is by 'linearising' a standard population dynamic model of the community by making the assumption that species are at equilibrium, thereby ignoring complications due to non-linear interaction terms between species. For example, given a deterministic community of n species, with the abundance of species i denoted by x_i, the dynamics of species i are given by:

$$\frac{dx_i}{dt}=\sum_{j=1}^{n} f_{ij}(x_i)x_j$$

where $f_{ij}(x_i)$ is a function defining the (potentially complex) interaction between species j and the focal species i. By assuming the species is at equilibrium (x_i^*) we can approximate this function using a Taylor series:

$$f_{ij}(x_i^*+h_i)=f_{ij}(x_i^*)+h_i f_{ij}'(x_i^*)+\frac{h_i^2}{2!}f_{ij}''(x_i^*)+\ldots$$

where $h_i=x_i - x_i^*$ represents a small departure of species i from equilibrium. Providing h is sufficiently small, we can neglect all but the first term:

$$f_{ij}(x_i^*+h_i) \approx f_{ij}(x_i^*)$$

allowing us to write the linearised system of equations as:

$$\frac{d\mathbf{x}}{dt}=\mathbf{Ax} \tag{4}$$

where $\mathbf{x}$ is a vector of species abundances and $\mathbf{A}$ is a *community matrix* with elements $a_{ij}=f_{ij}(x_i^*)$, describing the net interaction strength between species j and i. By linearising the dynamics of the community in this way, this approach reduces the number of parameters in the model (for example, by ignoring the shape of a predator's functional response).

Once the linearised model has been constructed these composite interaction terms (the a_{ij}) can be estimated either from knowledge of population sizes and energy conversion efficiencies for each interaction link (Raffaelli 2002), or from large-scale community perturbation experiments (Bender *et al.* 1984, Yodzis 1995, Schmitz 1997). A particularly useful approach is to construct models that contain various hypothesised interactions, parameterised as far as possible from logistically manageable single- or few-species lab and field experiments, and then use them to predict the response of various species to a perturbation (Schmitz 1997). In this way the perturbation is used to test the validity of the hypothesised interactions within the model.

Schmitz (ibid.) performed such an analysis of a simple grassland community comprising a single herbivore species (grasshopper), four competing plant species and the essential nutrient nitrogen. Using a hypothesised model of the system, parameterised from small-scale experiments, the community matrix ($\mathbf{A}$) of the ecosystem was constructed. Each element of the negative inverse of this matrix, $-[\mathbf{A}^{-1}]_{ij}$, then gave the expected net change in equilibrial density of each species before and after the perturbation, taking into account both direct and indirect interspecific effects; if $-[\mathbf{A}^{-1}]_{ij} > 0$ then perturbation of species j is predicted to result in an increase in the target species i, if $-[\mathbf{A}^{-1}]_{ij} < 0$ then perturbation of j should cause a reduction in the abundance of species i (Yodzis 1988, Schmitz 1997). Having predicted the responses of each species to perturbations involving the addition of either nitrogen or herbivores to the community, Schmitz conducted the necessary field experiments to test the predictions. The results of the experiments were remarkably consistent with the predictions, at least qualitatively, with each species responding in the predicted direction to each perturbation. However, the magnitudes of the predicted responses were generally poor; typically the model under-predicted the responses to the nitrogen perturbation and over-predicted the responses to the herbivore perturbation.

Related approaches include 'path analysis', which uses statistical methods such as multiple regression to explore the direct and indirect interactions between state variables in a community (Wootton 1994a, Wootton 1994b), and the analysis of 'trophic loops' (Yodzis 1989, Neutel *et al.* 2002), which are closed chains of interactions between adjacent species in a food chain. It is also interesting that Equation 4 has the same form as a continuous time linear Markov model of ecosystem states (e.g. the species present at a point in space). Over a finite time interval such a model is characterised by a matrix of transition probabilities among states. The resulting discrete-time Markov models (Waggoner and Stephens 1970, Wootton 2001) have been shown to make good predictions of the response of some communities to perturbations (Wootton 2001). The correspondence to Equation 4 suggests that this could arise either because a Markov model is a good mechanistic description of the system, or because it is a linearisation close to equilibrium of a more complicated non-linear system. The primary attractive feature of all these models is that because they are linear, estimation of parameters is easier than for non-linear models. Hence, the elements of the community or transition matrices are (relatively) easily derived from empirical studies of the community. However, these approaches are typically only reliable for relatively small perturbations near the equilibrium state, and predictions are likely to be inaccurate for larger perturbations. Nevertheless, the results of Schmitz (1997) and others are encouraging, suggesting that small-scale experiments and the appropriate modelling approach can be used to predict the direction and statistical significance of perturbations to (admittedly, simplified) ecosystems. The application of these approaches to more complex ecosystems remains an open question.

Aggregating state variables

Attempts have been made to simplify ecosystem models by aggregating species into functional groups or 'guilds', thereby reducing the dimensionality of the models (Yodzis 1988, Hawkins and MacMahon 1989, Simberloff and Dayan 1991, Yodzis and Winemiller 1999). Simply combining two species into a single functional group can reduce the number of interaction parameters that need to be estimated. In this case the general framework presented in Equation 1 can be used to describe the dynamics of functional group i. In the extreme case, it might be sufficient to simply predict the responses of trophic levels as a whole, rather than individual species (Abrams 1996).

Clearly however, the effectiveness of this approach depends on how the functional groups are defined and how few groups can be used without losing any essential components of the system (Yodzis 1988, Yodzis and Winemiller 1999). As a general rule it has been suggested that it is preferable to move away from taxonomic-based views of communities (i.e. species identities) and towards a functional- (or trait-) based view (McGill *et al.* 2006). Different methods of defining functional groups have been proposed, ranging from the fairly subjective to the reasonably objective. As an example of the latter, multivariate approaches may be used to define species groups (Hawkins and MacMahon 1989, Simberloff and Dayan 1991, Diaz and Cabido 1997). First, individual species are characterised in terms of their key traits or functional role in the ecosystem (e.g. body size, metabolic rates, feeding preferences, prey species consumed etc.). Multivariate statistics can then be applied to the distribution of trait values between species to determine 'clusters' (i.e. species that occur close together within trait space) that may be used to place similar species into groups (Yodzis and Winemiller 1999). However, although these species groupings are determined through statistical methods, there is still some degree of subjectivity; the threshold degree of similarity that creates the functional groups is ultimately an arbitrary decision. In general though, whatever method is used, groups should be constructed so that traits vary more between groups than within (McGill *et al.* 2006).

An obvious problem to the aggregation approach is how to determine the optimal level of complexity. One common approach is to compare the predictions of models of differing complexity with each other, rather than with data. That is, the predictions of simplified models with increasing degrees of aggregation are compared against those of the more complex, baseline model (Levin *et al.* 1997, Yodzis and Winemiller 1999, Fulton *et al.* 2003). However, previous analyses have produced conflicting findings, with optimal complexity ranging from very simple to very complex, depending on the system being analysed. For example, Ludwig and Walters (1981, Walters and Ludwig 1981) showed that a highly aggregated model could perform better than more complex ones, possibly due to the propagation of errors within more complex models. Typically though, models of intermediate complexity tend to be optimal (Fulton *et al.* 2003 and references therein). Hence, highly complex models are unlikely to be optimal – the low signal-to-noise ratios inherent in ecological data mean that a large model with inaccurate parameter values will not be any better than a simpler model with more precise estimates (Silvert 1981).

However, either when comparing models with each other, or models with empirical data, there is a fundamental problem of evaluating the performance of models with different levels of aggregation. That is, what outputs of the two models can be meaningfully compared? This problem can be expressed formally as follows. Consider an original model $\boldsymbol{M}_0$ for data $\mathbf{Y}^{\mathrm{T}}$ and an aggregated model $\boldsymbol{M}_1$ for data $\mathbf{Z}^{\mathrm{T}}$ (obtained by aggregating the original data). Instead of comparing two likelihoods $L(\theta_0;\mathbf{Y}^{\mathrm{T}}, \boldsymbol{M}_0)$ and $L(\theta_1;\mathbf{Y}^{\mathrm{T}}, \boldsymbol{M}_1)$ calculated from the same data, we are comparing $L(\theta_0;\mathbf{Y}^{\mathrm{T}}, \boldsymbol{M}_0)$ and $L(\theta_1;\mathbf{Z}^{\mathrm{T}}, \boldsymbol{M}_1)$. Because the likelihood functions are based on different data, we have no theory to tell us which model is better.

Aggregating parameters

An alternative, but related, approach that overcomes the problem of comparing different likelihoods is to identify groups of species with similar taxonomic or functional properties, and to assign the same parameters to all the members of this group. This approach is not as widely used as the method of aggregating state variables, but seems preferable because of its statistical properties. For example, suppose that two species i and k are hypothesised to belong to the same group, and that we are using a Lotka–Volterra model (e.g. Equation

3) in which the parameters associated with species i are a growth rate a_i and a set of interaction coefficients α_{ij}, where $j=1\ldots m$. Then in our aggregated model $\boldsymbol{M}_1$, we set $\alpha_{ij}=\alpha_{kj}$ j for all values of j and $a_i=a_k$. We are hypothesising here that species i and k are *demographically equivalent*. The idea of demographic equivalence has found recent prominence due to its being used extensively in the neutral models of species abundance distributions proposed by Hubbell (2001). In these models, Hubbell assumes that all individuals are demographically equivalent. However, if we are interested in the details of community dynamics rather than broad patterns of species abundance distribution, it may make more sense to consider less drastic simplifications in which only some groups of species are demographically equivalent (Pueyo *et al.* 2007).

In principle, unlike when aggregating state variables, it is easy to test whether a given simplification is justified. If our original model $\mathbf{M_0}$ had m species and a p dimensional parameter vector $\boldsymbol{\theta}_0$ estimated from the data, a new model $\boldsymbol{M}_1$ in which we assume that two species i and k are demographically equivalent will have a $p-(m+1)$ dimensional parameter vector $\boldsymbol{\theta}_1$ estimated from the data. The likelihoods $L(\boldsymbol{\theta}_0;\mathbf{Y}^{\mathrm{T}},\boldsymbol{M}_0)$ and $L(\boldsymbol{\theta}_1;\mathbf{Y}^{\mathrm{T}},\boldsymbol{M}_1)$ are calculated from the same data, and we can therefore use likelihood ratio tests to determine whether the simplification is acceptable. Aggregating parameters rather than state variables in this way will not give us a model with a new kind of dynamics. What it will give us is a way of comparing the fit of a simplified model and a more complicated model to the same data, which cannot be done if we aggregate state variables.

Isolated models for components of a complex community

A common simplification is to model the dynamics of some important subset of species (for example, those of economic or conservation importance, and any that are thought to have strong biological effects on the species of interest). At the extreme, we might construct an isolated single-species model. One well-known example of this approach is multispecies virtual population analysis (MSVPA), in which only the species of interest (e.g. economically important fish species, in the case of fisheries management) are modelled and all other components of the ecosystem are not explicitly included (Magnusson 1995, Livingston and Jurado-Molina 2000). Effectively these approaches simplify Equations 1 and 2, leading to:

$$dx_i=[f_i(x_i)+g_{ii}x_i^2]dt+h_i(x_i)dW_i(t)$$
$$y_i(t)=\phi_i(x_i(t))$$

The effects of temporal variability in other species can then only be absorbed into the noise function $h_i(.)$, which must be estimated from the data. *A priori*, we might expect an isolated model to work reasonably well for generalist consumers which are only weakly coupled to the dynamics of any individual prey species (Murdoch *et al.* 2002), whereas such models might work less well for species whose dynamics are strongly influenced by those of a few other species. However, if a species is strongly influenced by a resource species, but has little reciprocal influence on resource abundance, we could use the resource species as an environmental driver, rather than modelling its dynamics explicitly. This method has previously been used to investigate the impact of sandeel fisheries on kittiwakes (*Rissa tridactyla*), in which breeding success and survival were functions of sea surface temperature and the presence of a fishery effect (Frederiksen *et al.* 2004). Both these variables were assumed to influence the abundance of sandeels (*Ammodytes marinus*), which are the main prey species for kittiwakes in this region. In effect, Frederiksen *et al.* were treating sandeel abundance as an environmental driver and could then use their model to predict the possible responses of kittiwakes to future changes in fishery practices and sea surface temperature.

Single-species analyses are widely used in conservation biology, and reflect the application of population-level modelling to the study of natural ecosystems. A popular approach is to estimate multiple sets of parameters for an age- or stage-structured population model from several different years of data, and use these parameter sets to obtain an empirical estimate of variability in vital rates. We can then resample from these empirical estimates in a variety of ways to make predictions about the distribution of future population sizes, assuming that the distribution of sets of vital rates will remain constant (Caswell 2001, p. 415). That single-species models are moderately successful (Brook *et al.* 2000) suggests this can be a useful approach.

Evaluating isolated models requires us to define the set of species of interest in advance. Suppose that we define such a set S, and that we have a model $\mathbf{M_0}$ for set S. Suppose we also have a model $\mathbf{M_1}$ for a superset T which includes all the species in S and some additional ones. We can compare the quality of predictions from two models $\mathbf{M_0}$ and $\mathbf{M_1}$ for the subset of data $\mathbf{y}_S^T = \{\mathbf{y}_i^T : i \in S\}$, using the partial likelihoods $L(\theta_0; \mathbf{Y}_S^T, \mathbf{M_0})$ and $L(\theta_1; \mathbf{Y}_S^T, \mathbf{M_1})$. Because model $\mathbf{M_0}$ makes no predictions about the species in T that are not in S, we cannot compare $L(\theta_0; \mathbf{Y}_S^T, \mathbf{M_0})$ with $L(\theta_1; \mathbf{Y}_T^T, \mathbf{M_1})$.

Final thoughts and remaining challenges

Population, community and ecosystem ecology are vibrant fields of research, dating back decades. However, the links between them are not as obvious as they might be. The main discontinuities between these fields seem to arise from the scale of questions being addressed rather than any specific differences between the fields. Indeed, from an ecological perspective, the distinctions between an assemblage of interacting populations and a community, and between a community and an ecosystem, are somewhat arbitrary and difficult to determine – models of several interacting predator and prey species may correctly be regarded as being either population-level or community-level models or, if the state variables are measured in terms of carbon content rather than abundances, as ecosystem models. Therefore, the discontinuities really lie in the people doing the research rather than anything inherent in the assemblages themselves (cf. Raffaelli and Frid, this volume). This is primarily driven by the different scales of questions they are attempting to address; in general, population ecologists try to understand the mechanisms underlying variation in population abundances over time, community ecologists try to identify broad-scale characteristics of communities, and ecosystem ecologists concentrate on the processes underlying material or energy flows through ecosystems (DeAngelis 1992, Berlow *et al.* 2004).

Although these approaches differ in terms of their viewpoint of a species assemblage, they should be reconcilable. Indeed, given the current rates of biodiversity loss and the real threats of global climate change there is a pressing need to integrate these approaches to understand the relationships between environmental stressors, species loss, biodiversity and ecosystem processes (Jones and Lawton 1995, Loreau 2000). In particular, predicting the response of ecosystems to anthropogenic perturbations is crucial, and recent studies highlighting the reciprocal, beneficial role of biodiversity for humans emphasise this need even more (Diaz *et al.* 2006, Thuiller 2007).

To achieve such unity we must face a number of key challenges. A major obstacle to our progress is determining the role empirical data can play in driving model development, providing parameter estimates and testing and validating model predictions. This is effectively a question of scale, in terms of what is the appropriate scale at which to aim the model that will allow viable comparison with empirical data. As pointed out by Levin *et al.* (1997) it is not reasonable to expect even the most detailed individual-based models to accurately predict the location (or any other property) of every individual in the population; only aggregate properties can be expected to be reliably predicted over fairly broad spatial or temporal scales. This in turn raises the question: if predictions are only required (or useful, or feasible) at a broad scale, how much fine-scale detail needs to be included in the model to provide these predictions? More detailed work on how to formally develop and evaluate aggregated ecosystem models is needed to determine whether there are any general rules regarding the appropriate level of complexity for an ecosystem model.

Following on from this, increases in computing power are frequently praised for allowing us to develop more detailed models – but is this really a good thing? It appears that we are no longer limited by computing power, but by our ability to collect and analyse empirical data. Indeed, computing power has long since ceased to be an issue – our ability to collect the appropriate empirical data was exceeded years ago by the complexity of models that we could develop. So perhaps our attention should turn away from developing increasingly detailed models and move towards developing better frameworks for analysing experimental or observational data that can be used to drive model development. A first step in this direction would be for empiricists and theoreticians to collectively determine what exactly is meant by 'interaction' strength (for example, the g_{ij} terms in Equation 1, which have a clear biological and mathematical meaning) and collect data and design models that use the same meaning (de Ruiter *et al.* 1995, Berlow *et al.* 2004). Ideally such measures of interaction strength would be relevant at both the population and ecosystem levels (e.g. the energy flux-based measures of interaction strength suggested by Rooney *et al.* (2006), that relate to ecosystem-level turnover rates).

Finally, it is essential to incorporate one key component into ecosystem models; human activity (Crowder *et al.* 1996). There is major research effort currently under way to predict the impact of climate change on ecosystems, with attempts being made to predict temperature and precipitation changes over the coming years and then using these predictions to assess their impact at the population level (in terms of species abundances and distributions), the community level (in terms of biodiversity) and the ecosystem level (in terms of productivity and carbon and nitrogen flow). These models are frequently phenomenological, for example by relating current species distributions to environmental characteristics and then extrapolating in response to predicted future climate scenarios (Scott and Poynter 1991, Rogers and Randolph 1993, Jeffree and Jeffree 1996). However, this approach ignores many of the subtle, indirect effects within natural ecosystems, making such extrapolations hazardous (Davis *et al.* 1998). More mechanistic models may provide a better approach, allowing future scenarios to be modelled based on a firm understanding of the response of biological processes to temperature, humidity etc. Ultimately, combining population, community and ecosystem ecology and placing humans within the resulting framework is a major challenge facing ecologists in the twenty-first century. It is only through such an approach that we can hope to develop policies that lead to the sustainable management of natural ecosystems.

Acknowledgements

We are grateful to Rachel Bearon and David Montagnes for discussions, and to Jose Montoya, two anonymous referees, Dave Raffaelli, Chris Frid, and Odette Paramor for helpful suggestions.

References

Abrams, P. 1996. Dynamics and interactions in food webs with adaptive foragers. In: *Food webs: integration of patterns and dynamics* (Ed. by Polis, G. and Winemiller, K. O.), pp. 113–21: Chapman and Hall, New York.

Allen, J. I., Blackford, J. C. and Radford, P. J. 1998. An 1-D vertically resolved modelling study of the ecosystem dynamics of the middle and southern Adriatic Sea. *Journal of Marine Systems*, **18**, 265–86.

Allen, J. I., Somerfield, P. J. and Siddorn, J. 2002. Primary and bacterial production in the Mediterranean Sea: a modelling study. *Journal of Marine Systems*, **33**, 473–95.

Baretta, J. W., Ebenhoh, W. and Ruardij, P. 1995. The European Regional Seas

Ecosystem Model, a complex marine ecosystem model. *Netherlands Journal of Sea Research*, **33**, 233–46.

Barettabekker, J.G., Baretta, J.W. and Rasmussen, E.K. 1995. The microbial food-web in the European Regional Seas Ecosystem Model. *Netherlands Journal of Sea Research*, **33**, 363–79.

Bender, E.A., Case, T.J. and Gilpin, M.E. 1984. Perturbation experiments in community ecology: theory and practice. *Ecology*, **65**, 1–13.

Berlow, E.L., Neutel, A.M., Cohen, J.E., de Ruiter, P.C., Ebenman, B., Emmerson, M., Fox, J.W., Jansen, V.A.A., Jones, J.I., Kokkoris, G.D., Logofet, D.O., McKane, A.J., Montoya, J.M. and Petchey, O. 2004. Interaction strengths in food webs: issues and opportunities. *Journal of Animal Ecology*, **73**, 585–98.

Bickel, P.J. and Doksum, K.A. 2001. *Mathematical statistics, Volume 1: Basic ideas and selected topics*. Prentice-Hall, New Jersey. 2nd Edition.

Blackford, J.C., Allen, J.I. and Gilbert, F.J. 2004. Ecosystem dynamics at six contrasting sites: a generic modelling study. *Journal of Marine Systems*, **52**, 191–215.

Blackford, J.C. and Burkill, P.H. 2002. Planktonic community structure and carbon cycling in the Arabian Sea as a result of monsoonal forcing: the application of a generic model. *Journal of Marine Systems*, **36**, 239–67.

Bozdogan, H. 1987. Model Selection and Akaike Information Criterion (AIC): the general theory and its analytical extensions. *Psychometrika*, **52**, 345–70.

Brook, B.W., O'Grady, J.J., Chapman, A.P., Burgman, M.A., Akcakaya, H.R. and Frankham, R. 2000. Predictive accuracy of population viability analysis in conservation biology. *Nature*, **404**, 385–7.

Buckland, S.T., Newman, K.B., Thomas, L. and Koesters, N.B. 2004. State-space models for the dynamics of wild animal populations. *Ecological Modelling*, **171**, 157–75.

Caswell, H. 2001. *Matrix population models: Construction, analysis and interpretation*. Sinauer Inc., Sunderland, Massachusetts. 2nd Edition.

Christensen, V. and Pauly, D. 1992. ECOPATH II: a software for balancing steady state ecosystem models and calculating network characteristics. *Ecological Modelling*, **61**, 169–85.

Christensen, V. and Walters, C.J. 2004. ECOPATH with ECOSIM: methods, capabilities and limitations. *Ecological Modelling*, **172**, 109–39.

Crowder, A.A., Smol, J.P., Dalrymple, R., Gilbert, R., Mathers, A. and Price, J. 1996. Rates of natural and anthropogenic change in shoreline habitats in the Kingston Basin, Lake Ontario. *Canadian Journal of Fisheries and Aquatic Sciences*, **53**, 121–35.

Davis, A.J., Jenkinson, L.S., Lawton, J.H., Shorrocks, B. and Wood, S. 1998. Making mistakes when predicting shifts in species range in response to global warming. *Nature*, **391**, 783–6.

de Ruiter, P.C., Neutel, A.M. and Moore, J.C. 1995. Energetics, patterns of interaction strengths, and stability in real ecosystems. *Science*, **269**, 1257–60.

DeAngelis, D.L. 1992. *Dynamics of nutrient cycling and foodwebs*. Chapman and Hall, London.

Diaz, S. and Cabido, M. 1997. Plant functional types and ecosystem function in relation to global change. *Journal of Vegetation Science*, **8**, 463–74.

Diaz, S., Fargione, J., Chapin, F.S. and Tilman, D. 2006. Biodiversity loss threatens human well-being. *Plos Biology*, **4**, 1300–5.

Duffy, J.E. 2002. Biodiversity and ecosystem function: the consumer connection. *Oikos*, **99**, 201–19.

Ebenman, B. and Jonsson, T. 2005. Using community viability analysis to identify fragile systems and keystone species. *Trends in Ecology & Evolution*, **20**, 568–75.

Elton, C. S. 1958. *The ecology of invasions by animals and plants*. Methuen and Co., London.

Felsenstein, J. 2001. Taking variation of evolutionary rates between sites into account in inferring phylogenies. *Journal of Molecular Evolution*, **53**, 447–55.

Frederiksen, M., Wanless, S., Rothery, P. and Wilson, L. J. 2004. The role of industrial fisheries and oceanographic change in the decline of North Sea black-legged kittiwakes. *Journal of Applied Ecology*, **41**, 1129–39.

Fulton, E. A., Smith, A. D. M. and Johnson, C. R. 2003. Effect of complexity on marine ecosystem models. *Marine Ecology Progress Series*, **253**, 1–16.

Fulton, E. A. 2004. Biogeochemical marine ecosystem models I: IGBEM – a model of marine bay ecosystems. *Ecological Modelling*, **174**, 267–307.

Haputhantri, S. S. K., Villanueva, M. C. S. and Moreau, J. 2008. Trophic interactions in the coastal ecosystem of Sri Lanka: An ECOPATH preliminary approach. *Estuarine Coastal and Shelf Science*, **76**, 304–18.

Harrison, G. W. 1995. Comparing predator-prey models to Luckinbill's experiment with *Didinium* and *Paramecium*. *Ecology*, **76**, 357–74.

Harvey, A. C. 1989. *Forecasting, structural time series models and the Kalman filter*. Cambridge University Press, Cambridge.

Hawkins, C. P. and MacMahon, J. A. 1989. Guilds: the multiple meanings of a concept. *Annual Review of Entomology*, **34**, 423–51.

Heymans, J. J. and Baird, D. 2000. Network analysis of the northern Benguela ecosystem by means of NETWRK and ECOPATH. *Ecological Modelling*, **131**, 97–119.

Higham, D. J. 2001. An algorithmic introduction to numerical simulation of stochastic differential equations. *Siam Review*, **43**, 525–46.

Hubbell, S. P. 2001. *The unified neutral theory of biodiversity and biogeography*. Princeton University Press, Princeton.

Jansen, V. A. A. and Kokkoris, G. D. 2003. Complexity and stability revisited. *Ecology Letters*, **6**, 498–502.

Jeffree, C. E. and Jeffree, E. P. 1996. Redistribution of the potential geographical ranges of Mistletoe and Colorado Beetle in Europe in response to the temperature component of climate change. *Functional Ecology*, **10**, 562–77.

Jones, C. G. and Lawton, J. H. 1995. *Linking species and ecosystems*. Chapman and Hall, New York.

Kendall, B. E., Briggs, C. J., Murdoch, W. W., Turchin, P., Ellner, S. P., McCauley, E., Nisbet, R. M. and Wood, S. N. 1999. Why do populations cycle? A synthesis of statistical and mechanistic modeling approaches. *Ecology*, **80**, 1789–805.

Levin, S. A., Grenfell, B., Hastings, A. and Perelson, A. S. 1997. Mathematical and computational challenges in population biology and ecosystems science. *Science*, **275**, 334–43.

Livingston, P. A. and Jurado-Molina, J. 2000. A multispecies virtual population analysis of the eastern Bering Sea. *ICES Journal of Marine Science*, **57**, 294–9.

Loreau, M. 2000. Biodiversity and ecosystem functioning: recent theoretical advances. *Oikos*, **91**, 3–17.

Ludwig, D. and Walters, C. J. 1981. Measurement errors and uncertainty in parameter estimates for stock and recruitment. *Canadian Journal of Fisheries and Aquatic Sciences*, **38**, 711–20.

MacArthur, R. 1955. Fluctuations of animal populations, and a measure of community stability. *Ecology*, **36**, 533–6.

Magnusson, K. G. 1995. An overview of the multispecies VPA – theory and applications. *Reviews in Fish Biology and Fisheries*, **5**, 195–212.

May, R. M. 1974. *Stability and complexity in model ecosystems*. Princeton University Press Princeton.

McCann, K., Hastings, A. and Huxel, G.R. 1998. Weak trophic interactions and the balance of nature. *Nature*, **395**, 794–8.

McGill, B.J., Enquist, B.J., Weiher, E. and Westoby, M. 2006. Rebuilding community ecology from functional traits. *Trends in Ecology & Evolution*, **21**, 178–85.

Montoya, J.M., Rodriguez, M.A. and Hawkins, B.A. 2003. Food web complexity and higher-level ecosystem services. *Ecology Letters*, **6**, 587–93.

Murdoch, W.W., Kendall, B.E., Nisbet, R.M., Briggs, C.J., McCauley, E. and Bolser, R. 2002. Single-species models for many-species food webs. *Nature*, **417**, 541–3.

Neumann, T. 2000. Towards a 3D-ecosystem model of the Baltic Sea. *Journal of Marine Systems*, **25**, 405–19.

Neutel, A.M., Heesterbeek, J.A.P. and de Ruiter, P.C. 2002. Stability in real food webs: Weak links in long loops. *Science*, **296**, 1120–3.

Pauly, D., Christensen, V. and Walters, C. 2000. ECOPATH, ECOSIM, and ECOSPACE as tools for evaluating ecosystem impact of fisheries. *ICES Journal of Marine Science*, **57**, 697–706.

Petihakis, G., Smith, C.J., Triantafyllou, G., Sourlantzis, G., Papadopoulou, K.N., Pollani, A. and Korres, G. 2007. Scenario testing of fisheries management strategies using a high resolution ERSEM-POM ecosystem model. *ICES Journal of Marine Science*, **64**, 1627–40.

Pimm, S.L. 1984. The complexity and stability of ecosystems. *Nature*, **307**, 321–6.

Proulx, S.R., Promislow, D.E.L. and Phillips, P.C. 2005. Network thinking in ecology and evolution. *Trends in Ecology and Evolution*, **20**, 345–53.

Pueyo, S., He, F. and Zillio, T. 2007. The maximum entropy formalism and the idiosyncratic theory of biodiversity. *Ecology Letters*, **10**, 1017–28.

Raffaelli, D. 2002. Ecology: From Elton to mathematics and back again. *Science*, **296**, 1035–6.

Rogers, D.J. and Randolph, S.E. 1993. Distribution of tsetse and ticks in Africa: past, present and future. *Parasitology Today*, **9**, 266–71.

Rooney, N., McCann, K., Gellner, G. and Moore, J.C. 2006. Structural asymmetry and the stability of diverse food webs. *Nature*, **442**, 265–9.

Schmitz, O.J. 1997. Press perturbations and the predictability of ecological interactions in a food web. *Ecology*, **78**, 55–69.

Scott, D. and Poynter, M. 1991. Upper temperature limits for trout in New Zealand and climate change. *Hydrobiologia*, **222**, 147–51.

Silvert, W.L. 1981. Principles of ecosystem modelling. In: *Analysis of marine ecosystems* (Ed. by Longhurst, A.R.), pp.651–76: Academic Press, New York.

Simberloff, D. and Dayan, T. 1991. The guild concept and the structure of ecological communities. *Annual Review of Ecology and Systematics*, **22**, 115–43.

Thuiller, W. 2007. Biodiversity: Climate change and the ecologist. *Nature*, **448**, 550–2.

Vichi, M., Pinardi, N., Zavatarelli, M., Matteucci, G., Marcaccio, M., Bergamini, M.C. and Frascari, F. 1998. One-dimensional ecosystem model tests in the Po Prodelta area (Northern Adriatic Sea). *Environmental Modelling & Software*, **13**, 471–81.

Waggoner, P.E. and Stephens, G.R. 1970. Transition probabilities for a forest. *Nature*, **225**, 1160.

Walters, C.J. and Ludwig, D. 1981. Effects of measurement errors on the assessment of stock-recruitment relationships. *Canadian Journal of Fisheries and Aquatic Sciences*, **38**, 704–10.

Whelan, S. and Goldman, N. 2001. A general empirical model of protein evolution derived from multiple protein families using a maximum-likelihood approach. *Molecular Biology and Evolution*, **18**, 691–9.

Wootton, J.T. 1994a. The nature and consequences of indirect effects in ecological communities. *Annual Review of Ecology and Systematics*, **25**, 443–66.

Wootton, J.T. 1994b. Predicting direct and indirect effects: an integrated approach using experiments and path analysis. *Ecology*, **75**, 151–65.

Wootton, J.T. 2001. Prediction in complex communities: Analysis of empirically derived Markov models. *Ecology*, **82**, 580–98.

Yodzis, P. 1988. The indeterminacy of ecological interactions as perceived through perturbation experiments. *Ecology*, **69**, 508–15.

Yodzis, P. 1989. *Introduction to theoretical ecology*. Harper and Row, New York.

Yodzis, P. 1995. Food webs and perturbation experiments: theory and practice. In: *Food webs: integration of patterns and dynamics* (Ed. by Polis, G. and Winemiller, K.O.), pp. 192–200: Chapman and Hall, New York.

Yodzis, P. and Winemiller, K.O. 1999. In search of operational trophospecies in a tropical aquatic food web. *Oikos*, **87**, 327–40.

CHAPTER THREE

Thermodynamic approaches to ecosystem behaviour: fundamental principles with case studies from forest succession and management

PAUL C. STOY

School of GeoSciences, University of Edinburgh

Introduction

Ecosystems and organisms must obey physical laws. This statement, perhaps due to its obviousness, is extremely powerful. It forms the basis of how we model systems, living or otherwise, to understand their dynamics and behaviour. Mass and energy must be conserved, but many physical configurations can satisfy the conservation of mass or energy. Ecosystems follow the *laws of thermodynamics*, and the ways in which ecosystems obey these laws determine their behaviour.

This chapter discusses how classic and contemporary ideas from physics (via thermodynamics) and statistics (via information theory) have influenced the study of ecology. After reviewing the history of the thermodynamic approach in biology, basic physical and statistical concepts are reviewed, and their practical application demonstrated, and debated, using case studies of temperate forest succession in the south-eastern United States and global forest management for atmospheric CO_2 mitigation after the Kyoto and Bali accords. Throughout, the different viewpoints of community ecology and ecosystem ecology are contrasted to place thermodynamic principles in a broader ecological context, and to explore ways to improve existing ecological theories.

Historical development and motivation

The thermodynamic approach to understanding biological systems was articulated most elegantly in a series of lectures by Erwin Schrödinger, recapitulated in a book entitled *What is Life* (1944). Schrödinger describes living systems as those that dissipate energy, or pass *entropy* to their surroundings,[1] to maintain an ordered state that is far from thermodynamic equilibrium. Equilibrium is a state in which no net energy is transferred either within or into a system, which means death to an organism. By dissipating energy, organisms contribute to

[1] The idea that living systems export disorder to maintain a state of relative order has been called *negentropy* and given a formal statistical definition as the excursion from a normal (Gaussian) distribution (Brillouin 1953).

Ecosystem Ecology: A New Synthesis, eds. David G. Raffaelli and Christopher L. J. Frid. Published by Cambridge University Press.

universal entropy despite being locally ordered (more ordered than their surroundings) and thereby embody the Second Law of Thermodynamics, discussed later. To maintain and develop the gradients through which energy and matter are transferred, organisms contain *information* that maintains their status and forms the basis for their future development (Schrödinger 1944, Ulanowicz 1986, Jørgensen *et al.* 2007). The molecular basis for heredity had not been discovered when Schrödinger gave his lectures, but it is intuitive that DNA represents this information in part, along with physical constraints on biological systems.

Ecosystems can also be described using this *holistic* approach. Like organisms, they maintain a state far from thermodynamic equilibrium by exploiting embodied information, and contribute to universal entropy by dissipating energy to maintain a local state of relative order. At the same time, ecosystems are comprised of multiple organisms, and this presents a conundrum for ecological modelling that recalls a debate from the early twentieth century (see Raffaelli and Frid, this volume) captured by the following questions: Are ecosystems best described (Gleason 1926) and modelled (Moorcroft *et al.* 2001, Clark 2003) as a collection of organisms? Or do community assemblages (Clements 1936) and ecosystems (Odum 1969) behave in a predictable fashion despite their complexity, such that modelling the whole system can accurately predict its behaviour?

Addressing these questions in part, and complementing Schrödinger's work from the point of view of ecological succession, is E.P. Odum's conceptual model, the 'Strategy of Ecosystem Development' (SED) (Odum 1969). The SED incorporates the role of community ecology in ecological succession by envisioning a shift in dominance between fast-growing '*r*-strategists' and slower growing *K*-strategist species as ecosystems mature (Figure 3.1) and lists a suite of ecosystem attributes that tend to change directionally over time (Figure 3.1), see also White *et al.*, this volume, Table 4.3.

One may argue that the SED is Clementsian (or teleological), as if the ecosystem has a strategy to develop itself in what Odum calls an 'orderly process'. But ecosystems are not sentient and the effects of individual actors on the system as a whole must be reconciled with the holistic viewpoint to understand *why* succession (usually) proceeds. Recent work has re-formulated the tendency toward the climax state in probabilistic terms: for a given event the chance of 'forward' progression toward climax is greater than that for one that reverts the ecosystem to a previous state (Jørgensen *et al.* 2007). Noting that ecosystem dynamics are *irreversible*, the ecosystem cannot step backwards to perfectly embody a previous state because its configuration has changed (Prigogine 1961). In brief, to understand ecosystem dynamics, we must combine insights from both thermodynamics and statistics to couple physical laws, biological processes and chance events. This is a challenging prospect, especially from

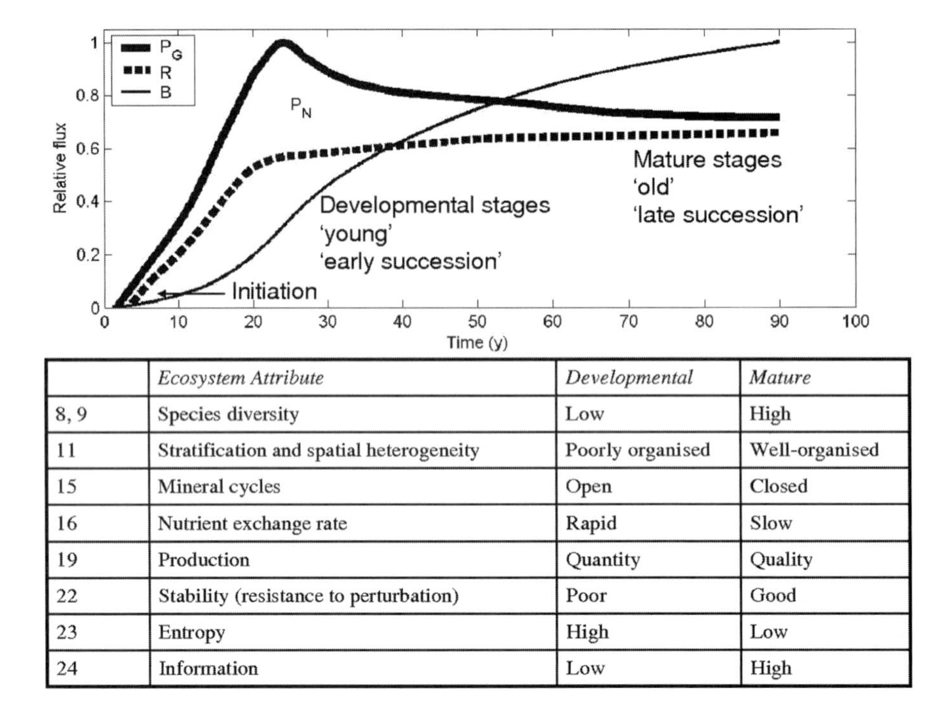

	Ecosystem Attribute	*Developmental*	*Mature*
8, 9	Species diversity	Low	High
11	Stratification and spatial heterogeneity	Poorly organised	Well-organised
15	Mineral cycles	Open	Closed
16	Nutrient exchange rate	Rapid	Slow
19	Production	Quantity	Quality
22	Stability (resistance to perturbation)	Poor	Good
23	Entropy	High	Low
24	Information	Low	High

Figure 3.1 (Upper) The 'Strategy of Ecosystem Development' (SED, Odum 1969) digitised and re-drawn. Gross primary productivity (P_G), respiration (R) and biomass (B) – the change over time according to the influence of ecosystem attributes on the physical environment, a selection of which are presented in the (lower) panel after Table 1 in Odum (1969). The difference between P_G and R is net primary productivity, P_N. As written, this figure only considers autotrophic biomass. Throughout the manuscript, the heterotrophic component of R will be added and the net ecosystem productivity (NEP, also called net ecosystem exchange, NEE) will replace P_N for full ecosystem carbon accounting. Numbers in left hand column are Odum's attributes; see text.

the applied perspective of ecosystem management, but the fundamental concepts are intuitive and follow easily from basic principles.

Definitions and terms

Systems science

The following definitions pertain to *open* systems; those that have a boundary, always user-defined, across which mass and energy are readily transferred. All ecosystems are open systems. Sometimes the boundaries that we draw around ecosystems are intuitive, such as the transition from a stream to its non-aquatic surroundings. Sometimes they are less intuitive, such as defining the effective riparian zone of near-stream vegetation that acts as an important interface for biogeochemical cycling (Naiman and Décamps 1997, Hedin *et al.* 1998). Defining system boundaries is required to understand system behaviour.[2]

[2] The interpretation and definition of ecosystem boundaries may be critical to policy and law. For example, the protection of US wetlands after the Clean Water Act of 1972 relies heavily

In contrast, the field of thermodynamics was largely developed with closed systems, such as engines, in mind. The following introduces thermodynamic principles from the open system perspective as applied to ecosystem ecology, and the basics of information theory as applied to thermodynamics.

Physical principles

The foremost physical concept is that of mass (m, SI units kilograms, kg), a conserved quantity in basic (Newtonian) physics.[3] A force (F, SI units Newtons, N, kg m s^{-2}) is required to accelerate m after Newton's Second Law of Motion

$$F = ma, \tag{1}$$

where a is acceleration. Another conserved quantity, energy (U, SI units Joules, N m, kg m^2 s^{-2}) is required to act against a force for some distance (d)

$$U = Fd = W; \tag{2}$$

this is also the definition of mechanical work (W), noting also that chemical bonds contain energy which living systems may be able to exploit. *Exergy* is then defined as the maximum amount of work that can be performed by the system that brings it into equilibrium with its surroundings. As ecosystems and organisms acquire mass and the chemical bonds that support it, their exergy increases.

Classical thermodynamics

The *First Law of Thermodynamics* states most simply that energy can be neither created nor destroyed (i.e. it is conserved, Mayer 1841 (in Lehinger 1971)), but can be transferred to different forms. The First Law gives us an equation for the change in U (dU) in an open system

$$dU = \delta Q - \delta W + U_{ex}. \tag{3}$$

dU is a function of work performed by the system on its surroundings (δW) and change in heat (δQ), acknowledging that energy can enter and leave the open boundaries of the system via U_{ex}.

One example of the First Law is the conversion of radiative to chemical energy via photosynthesis: energy is conserved and photosynthetic products can perform work when reduced chemical bonds are oxidised. Heat (Q) is created during the process (Equation 3). This is often referred to as 'waste heat' in colloquial terms.

Equation 3 can be expanded by noting that W can also be defined as the product of pressure and the change in volume $\delta W = pdV$ and that heat is defined as

on the definition of a wetland and its boundaries, in this case its hydrological connectivity with its surroundings.

[3] The conservation of mass applies to the laws of motion after Newton, not special relativity after Einstein, famously articulated by equating mass and energy, $E = mc^2$. Mass is conserved in as much as it *is* energy (and vice versa).

temperature times the change in *entropy* of a system, $\delta Q = TdS$. Equation 3 can then be rewritten:

$$dU = TdS - pdV + U_{ex}. \quad (4)$$

If exergy is the availability of a system to perform work (pdV), entropy is the term in the total energy balance that is unable to do work (TdS).

According to the *Second Law of Thermodynamics*, entropy in an isolated system always increases. Removing the last term on the right hand side of Equation 4, this is equivalent to saying that the ability of an isolated system to do work does not spontaneously increase, or a system cannot obtain perfect efficiency.

It is apparent that the First and Second Laws are difficult to separate if the change in heat is related to the change in entropy, which makes up part of the total energy balance (Kleidon, 2008). Any work performed by a system is accompanied by an increase in entropy after the Second Law; note the sign convention in Equation 4. The *Combined Law of Thermodynamics* is derived by simply rearranging Equation 4, adding the Second Law definition that the entropy of isolated systems increases:

$$dU - TdS + pdV - U_{ex} \leq 0 \quad (5)$$

The first three terms of the left-hand side of the equation equal the change in *Gibbs energy* ('*Gibbs free energy*') of the system.

Statistical thermodynamics (statistical mechanics)

The laws of thermodynamics were derived to describe macro-scale phenomena. It became clear after the discovery of atomic particles that these phenomena have a molecular basis, and the field of *statistical mechanics* was developed in conjunction with the prior description of irreversibility by Carnot. Molecular motion appears random at the macro scale, so quantifying macro-scale observations requires a probabilistic description of the microstates (Ω) that result in the observed macro-scale behaviour. Assuming the system is at equilibrium, the state that is observed is probably the one that has the highest probability of system microstates configurations. For an equilibrium system, entropy is defined as:

$$S = k_b \log \Omega \quad (6)$$

where k_b is the Boltzmann constant. Equation 6 is known as the Boltzmann equation, and quantifies the increase in entropy with an increasing number of molecular microstates.[4]

[4] Upon examining the molecular explanation of entropy, it was realised that there is a finite chance that order may spontaneously increase due to random fluctuations (the Fluctuation Theorem (Evans and Searles 1994, Wang *et al.* 2002)). The greater the scale of observation, the more likely the Second Law will hold. Therefore, the Second Law is a merely a statistical construct, further intertwining thermodynamics and statistics.

If the change in *distribution* of microstates is the quantity of interest considered, entropy can be expressed in a similar fashion, noting that for the molecular case, the distribution is discrete:

$$S = -k_b \sum_{i=1}^{n} p(x_i) \log p(x_i). \qquad (7)$$

If each state is equally likely, entropy is maximised. The Gibbs Algorithm for quantifying the role of the microstate ensemble on the macro scale is simply Equation 7 without the preceding sign and Boltzmann constant.

For the non-equilibrium case, in which the system develops over time, the Maximum Entropy (MaxEnt) school of thinking asserts that the Gibbs Algorithm also holds for the dynamic trajectory (path) taken by the microstates (Jaynes 2003). Jaynes (1957) introduced the term *path information entropy* (as discussed in Dewar 2003) to quantify entropy using:

$$S_I = -\sum p_\Gamma \log p_\Gamma \qquad (8)$$

where p_Γ is the microscopic path distribution, determined by constraints on the macroscopic system (Dewar 2003). The macro-scale observation probably results from the most likely distribution of paths.

The definitions of entropy or path entropy from Equations 6–8 may seem opaque, so a simple example with ecological implications may help. Consider a row of trees along a road. If the large-scale observation is that the row is formed of a single species, there is one possible 'microstate', a series of trees of one species. Each new observation adds no additional information: all the trees are the same. If the large-scale observation is that multiple tree species are present, multiple system configurations can give that result. There is a single distribution for the single-species case, and multiple distributions for the multi-species case. If the system is dynamic and the species composition and order change over time, the non-equilibrium case (Equation 8) is more appropriate but requires knowledge of how recruitment and succession result in growth trajectories. In all cases, more information is necessary to quantify the multi-species system, so there is a fundamental connection between entropy and information.

Information theory

Information can be quantified as the minimum amount of bits required to communicate a discrete idea, a bit being the amount of information necessary to make a binary decision. As opposed to energy, information is not conserved and can be readily lost, or, alternately, created. An intuitive example is erasing a computer disk. The ability of DNA to create proteins for growth or maintenance is lost shortly after the death of an organism. The loss of a keystone species changes the information content of an ecosystem by reorganising the flow of organisms, energy and nutrients.

Despite differences in conservation properties, information, like thermodynamics, can be quantified by its entropy:

$$H(X) = -\sum_{i=1}^{n} p(x_i) \log p(x_i) \quad (9)$$

where p is a probability density function (pdf). Note that, per the sign convention, information gain implies a decrease in $H(X)$. Equation 9 is the famous *Shannon Entropy* (Shannon 1948). The magnitude of information content is maximised with a uniform distribution (for a discrete distribution with 10 equal bins, $H(X) = -\sum_{i=1}^{N} p(x_i) \log p(x_i) = -10(.1 \times \log .1) = 2.303$, Figure 3.2 (left) and minimised with the Dirac delta function $H(X)=0$, Figure 3.2 (right)). Returning to the example of the trees, if a single species is modelled as one 'bin' of information, then $H(X)=0$. The Shannon entropy is the foundation for quantifying the amount of information shared (via, e.g. the mutual information content) or, critically, how information is processed and transferred by a system.

The information state of a system – or what we know about a system – can change. Relative changes in information can be quantified using, for example, the Kullback–Leibler divergence D_{KL}, which compares p with a resultant pdf or *a posteriori* distribution, q:

$$D_{KL} = \sum_{i=1}^{n} p(x_i) \log \frac{p(x_i)}{q(x_i)}, \quad (10)$$

The case of a continuous distribution is occasionally called the relative entropy (Jaynes 2003) or 'information gain' in Bayesian statistics:

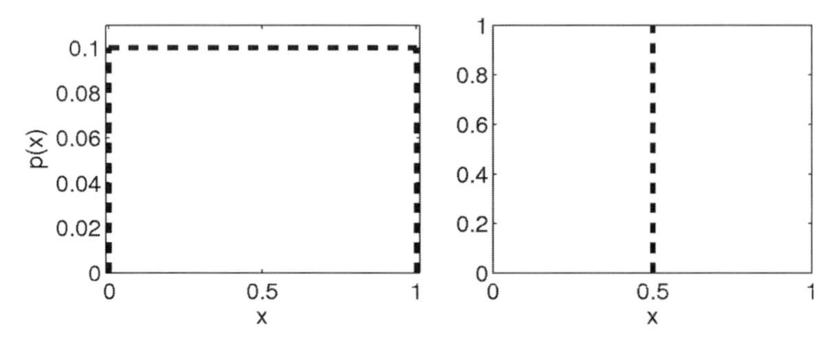

Figure 3.2 The Shannon entropy for two distributions with the maximum difference in information content, a uniform distribution (left) and a Dirac delta function (right). For a uniform distribution defined between $x=0$ and 1, $p(x)=0.1$ for all bins if $N=10$ bins. $H(X) = -\sum_{i=1}^{N} p(x_i) \log p(x_i) = -10(.1 \times \log .1) = 2.303$. The uniform distribution maximises the Shannon entropy; all other distributions with $N=10$ have lower H. For the case of the Dirac delta function, regardless of bin number, the probability is 1 at $x=0.5$ and the Shannon entropy is $H(X) = -\sum_{i=1}^{N} p(x_i) \log p(x_i) = -10(.1 \times \log 1) = 0$.

$$\int p(x)\log\frac{p(x)}{q(x)}dx \tag{11}$$

and analytical expressions for the Kullback–Liebler divergence can be written for different statistical distributions (e.g. Normal, Beta, Gamma). In other words, the distribution of information defines the information entropy, just as the distribution of molecular microstates defines the molecular entropy. For the case of ecosystems, the information entropy may refer to the distribution of species (the Shannon diversity is simply Equation 7 for the case of species), the distribution of chemical bonds that are oxidised in ecosystem respiration (Davidson and Janssens 2006), or whatever dynamic of the ecosystem the observer is interested in quantifying. As ecosystems conserve energy and contain information, thermodynamic and statistical concepts can be applied to interpret their behaviour.

Ecosystem applications

Changes in the thermodynamic state and information content of ecosystems are coupled: a change in the dissipation of energy implies a change in system structure that represents a change in system information. In ecosystems, these changes are often directional, and certainly irreversible (Prigogine 1961), and tend to result in ecosystem succession. Multiple theories and techniques have been developed to couple theory and ecosystem dynamics, often with a focus on ecological succession. Three examples are described next, then critiqued using a case study of temperate forest succession. Some key ideas are then extracted to describe the elements necessary to combine individual-based and ecosystem-based models for an improved description of ecosystems.

The Strategy of Ecosystem Development (SED)

The SED (Odum 1969) is a conceptual model of ecosystem development that envisions directional and predictable changes along ecological succession resulting from modifications of the physical environment by the ecological community. The SED has been used to interpret food webs, nutrient cycling, biological diversity and human ecology among other concepts. For simplicity, we will focus on its predictions of carbon cycling and ecosystem metabolism following the examples given in the original publication.

Some major elements of the SED are described in Figure 3.1 and the embedded table. Upon ecosystem establishment, fast-growing early successional species with a high rate of biomass acquisition dominate community composition. The photosynthesis/respiration (P/R) ratio is large and biomass (B) rapidly increases. This rapid rate of growth comes at the expense of sensitivity to environmental perturbations. This 'young nature' assemblage is eventually replaced by slower-growing species representative of an older or more mature ecosystem. The result is a stable ecosystem with large biomass and information content, but little additional biomass growth (i.e. the P/R ratio tends toward zero).

Ascendancy

Energy inputs, outputs, and the pathways that energy travels in a system can be quantified by the system's *ascendancy* (Ulanowicz 1986), which explicitly couples thermodynamic constraints with the information content of a system.

This derivation of ascendancy follows Ulanowicz and Jørgensen *et al.* (Jørgensen *et al.* 2007). Consider a case where a quantum of mass or energy[5] is transported within an ecosystem from ecosystem element i to j. The joint probability that the transfer (T) travels along the constrained direct pathway from ecosystem element i to j is T_{ij}/T. There is also a smaller probability that this transfer is along an indirect pathway, say $i \rightarrow k \rightarrow j$. The probability of a transfer leaving i to all ecosystem components directly connected to i (q), of which k is an element, is $\Sigma_q T_{iq}/T$. Likewise, the probability of the transfer entering j from connected elements r is $\Sigma_r T_{rj}/T$. The joint probability of the indirect transfer is then $\Sigma_q T_{iq} \Sigma_r T_{rj}/T^2$. As all transfers rely on the probability of moving from i to j, Ulanowicz (1986) applied the Boltzmann formula (Equation 7) to the difference between the unconstrained and constrained transfers for the case of all ecosystem transfers to derive an expression for ecosystem ascendancy (A):

$$A = \sum_{i,j} T_{ij} \log \frac{T_{ij} T}{\sum_k T_{kj} \sum_q T_{iq}}. \qquad (12)$$

All else being equal, a system with more states and flows has greater ascendancy. Similar statistical logic can be used to quantify other terms that describe the system such as the diversity of flows (similar in form to the Shannon entropy) and capacity for the system to develop further. Together, these terms can be used to describe not only an ecosystem's current state, but also its ability to change over time.

Ulanowicz (Ulanowicz 1980, 1986) derived a phenomenological principle based on ascendancy that is consistent with the observations of Odum (Odum 1969) and others regarding the directionality of ecosystem development: 'in the absence of major perturbations, ecosystems have a propensity to increase in ascendency'. In other words, as ecosystems develop, their structure becomes more complex and their information content and ascendancy increase. The box-models used in the derivation of ascendancy can be made to conserve mass and energy. In short, ascendancy combines thermodynamic laws and information theory (e.g. Equation 12) to describe ecosystems. The development of the ascendancy concept is progressing to quantify how (eco)system information is clustered, the sustainability of systems, and the amount of information missing from systems (Ulanowicz *et al.* in press).

[5] For example, mass via carbon compounds or energy via the chemical bonds that these compounds contain.

Ecological Law of Thermodynamics and Life as the Second Law

Sven Erik Jørgensen and others have combined the above ideas from thermodynamics and statistics in a postulate they call the *Ecological Law of Thermodynamics* (ELT): 'A system will attempt to utilise the flow of exergy to increase its exergy to move away from thermodynamic equilibrium; if more combinations and processes are offered (i.e. if more information is available) to utilise the exergy flow, the organisation that gives the highest exergy will be selected'(Jørgensen 1997, de Wit 2005, Jørgensen *et al.* 2007). This is similar to the postulate posed by Schneider and Kay (1994), who argued that 'ecosystems develop in ways which systematically increase their ability to degrade the incoming solar energy' by developing more complex structures, diversity and hierarchical levels to increase energy dissipation and flow.

The ELT can be viewed as an extension of thermodynamic laws for the case of ecosystems. If organisms are contributing to global entropy they must be acquiring exergy after Equation 4. There is no law from classical thermodynamics that states that entropy must be maximised (only that it tends to increase at macroscopic scales), but adherents of the maximum power principle have long recognised that biological systems tend toward maximum efficiency, arguing that this supports the Darwinian preservation of type (Lotka 1922a, b). Some authors such as Howard T. Odum have proposed that this biological incursion of thermodynamic principles is sufficiently general and supported to become the *Fourth Law of Thermodynamics*[6] (Odum 1994). The ELT formalises some of these ideas in a simple statement that applies thermodynamic concepts to ecological succession. The utility of the SED, ascendancy and the ELT are briefly discussed in the context of recent measurements of ecosystem metabolism, via biosphere–atmosphere carbon flux, along an ecosystem successional trajectory.

Critiques and a simple reformulation

All ecological theories have their advantages and disadvantages, may be overly abstract, oversimplified, not universally applicable, or too difficult to test. We may be able to extract ideas that can make incremental improvements for applied ecology by critiquing the theories outlined above. Temperate forest succession will be used as a case study to follow a major theme of the SED.

Critique of the Strategy of Ecosystem Development

The SED has stood the test of time, until recently, as an intuitive empirical and conceptual model of ecological succession that qualifies the change of an ecosystem based on the typical changes of the species that comprise it. Generations of ecologists are familiar with the SED because E.P. Odum wrote (or co-wrote)

[6] The Third Law of Thermodynamics states that entropy approaches zero as temperature approaches absolute zero.

many popular textbooks on ecology (e.g. Odum 1971). Despite the widespread use of the SED, it may oversimplify the story of ecosystem C uptake over time. The predictions of the carbon balance of forested ecosystems (Figure 3.1b in Odum (1969) were based on a single suite of temperate forest plantations (Kira and Shidei 1967). Very few actual data were used in the original creation of the SED, and it should be revisited using modern measurements.

For the case of temperate and boreal forest ecosystems, contemporary research has demonstrated that the coupling between carbon inputs via photosynthesis and outputs via ecosystem respiration (Reichstein *et al.* 2007) may result in substantial ecosystem C uptake in 'older' ecosystems (Baldocchi 2008, Luyssaert *et al.* 2008) despite predictions from the SED that older ecosystems are carbon neutral. In other words, net primary (or ecosystem) productivity may not tend toward zero as ecosystems become old (Knohl *et al.* 2003) because younger individuals take the place of older individuals (Moorcroft 2003) and the C balance of older individuals may have been underestimated in some cases (Carey *et al.* 2001, Röser *et al.* 2002). Soil carbon stocks may also increase in older ecosystems (Zhou *et al.* 2006), highlighting the critical role of the boundary in ecological science.

The SED considers the case of autotrophic biomass, but for full C accounting the net ecosystem productivity (NEP), or net ecosystem exchange (NEE) is more relevant as it describes the entire system, including heterotrophic respiration from the soil. Stoy *et al.* (2006b, 2008) examined the SED predictions using eighteen site-years of continuous eddy covariance measurements in three adjacent ecosystems that represent initial, fast-growing and mature stages of ecological development in the south-eastern United States. The eddy covariance technique measures, continuously and in a non-invasive manner, the biosphere–atmosphere CO_2 flux of ecosystems (Baldocchi *et al.* 2001). It was found that an older, late-successional hardwood forest (HW) and a younger, early-successional pine forest (PP) had identical NEE over five common years of measurement because of the sensitivity of the younger pine stand to drought and disturbance (Figure 3.3 Upper). In the context of the SED, the tendency of the older ecosystem to increase 'protection' against disturbance resulted in the same interannual ecosystem C uptake as in the younger ecosystem (Figure 3.3 Lower). The SED predicts that PP is more productive. Measurements demonstrated that this assumption held under ideal scenarios during periods less impacted by disturbances (Figure 3.3A); in other words periods when ecosystem structure was at 'steady state' and not perturbed by drought or disturbance (in this case an ice storm (McCarthy *et al.* 2006)).

At the same time, many forest measurements demonstrate a decrease in C uptake with ecosystem age (Magnani *et al.* 2000). As these studies are often from the field of forestry, they tend to involve stands with one or a few dominant species of a common age (i.e. a system with less information), and the dynamic

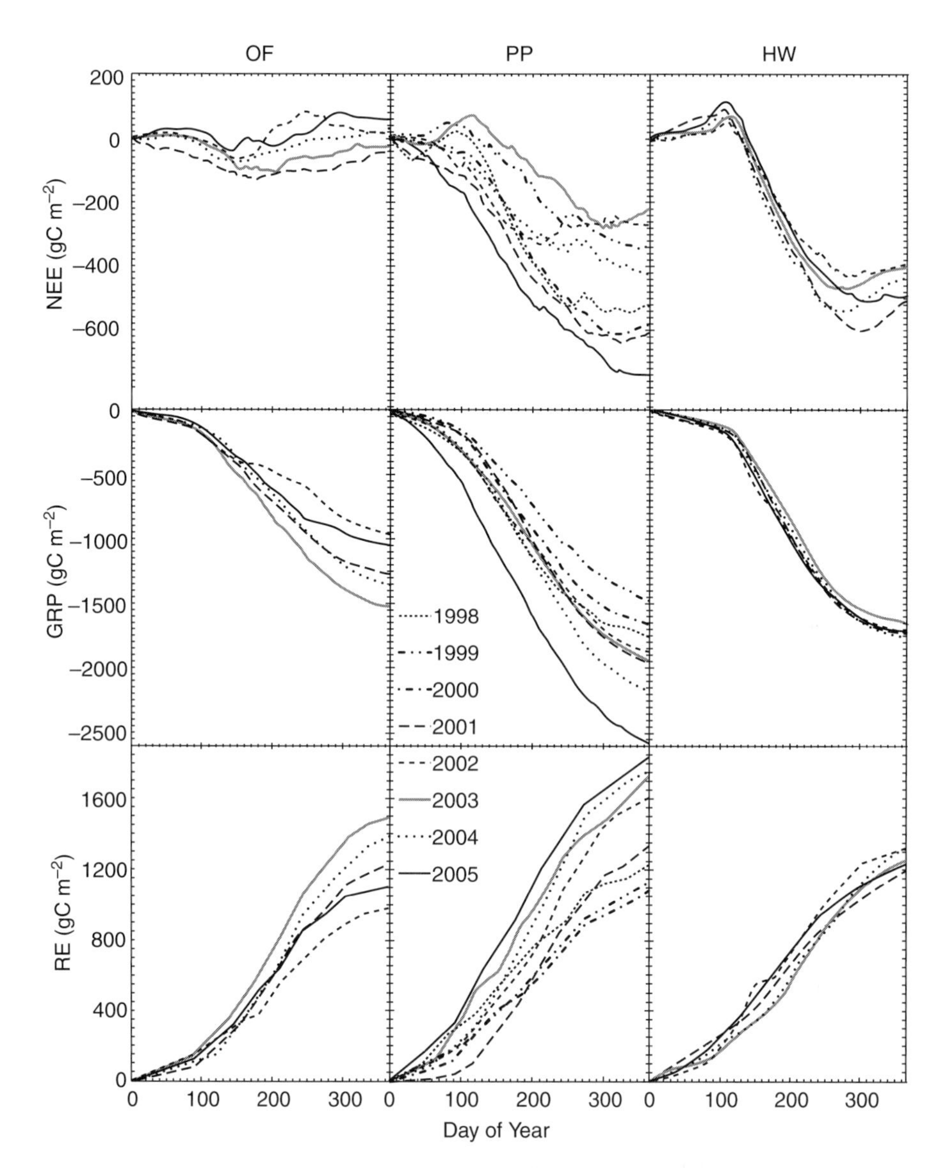

Figure 3.3 (Upper) The cumulative sum of net ecosystem exchange of carbon (NEE, upper panels), gross ecosystem productivity (GEP, middle panels) and ecosystem respiration (RE, lower panels) in old field (OF), planted pine (PP) and hardwood forest (HW) ecosystems in the Duke Forest, NC, after Stoy *et al.* (2008). (Lower) The interannual mean and variability (expressed as error bars) of observed GEP and RE are plotted as a function of approximate ecosystem age. Figure 1A of the 'Strategy of Ecosystem Development' (Odum, 1969) (thin lines) was digitised and rescaled to approximate the magnitude of GEP and RE observed by Stoy *et al.* (2006b, 2008). The abscissa was unchanged from Odum (1969). [Redrawn with permission from Blackwell Publishing.]

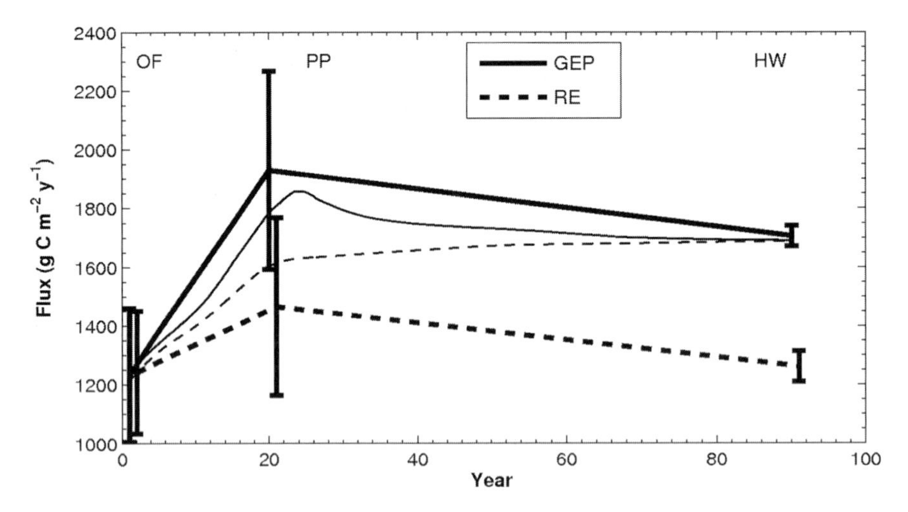

Figure 3.3 (cont.)

may reflect the life history of individual trees rather than more natural forest with multiple tree species of different sizes and ages. To expand on this point in the context of the SED, consider a case where an individual, rather than an ecosystem, embodies the SED C flux/time curve from Figure 1A in Odum (1969) (Figure 3.1 here). It is not unreasonable to assume that an individual exhibits fast growth early in its development and slower growth (or decline) with age, and that individuals progressively (and randomly) colonise an ecosystem as it ages (Figure 3.4 Upper). It is also relatively easy to incorporate trends in the life history of the hypothetical individuals from fast-growing r-type species to slower-growing K-type species (Odum, 1969) (Figure 3.4 Lower).

The simplistic, stochastic individual-based model in Figure 3.4 does not predict a substantial decline in biomass B with age, rather a close relationship between ecosystem carbon uptake (P) and loss (R) with both simulations. This relationship arises from the shape of the proposed individual life history curve. There is also a close formal coupling between P and R in different terrestrial ecosystems (Ekblad and Högberg 2001, Högberg *et al.* 2001), but P and R were not mathematically coupled in this example.

The hypothetical individual-based model does not consider the filling of canopy space, recruitment, actual species composition, real tree life history or mortality, but it is clear that models inspired by individual or community dynamics may arrive at very different conclusions than the whole-system perspective regarding C uptake with ecosystem age. For example, Moorcroft *et al.* (2001) demonstrated that an individual-based stochastic canopy gap model (the 'ecosystem demography' model, ED) was able to capture the observed increase in above-ground biomass with age since establishment in tropical rainforest ecosystems (see Figure 4A ibid.). Size-structure models predicted

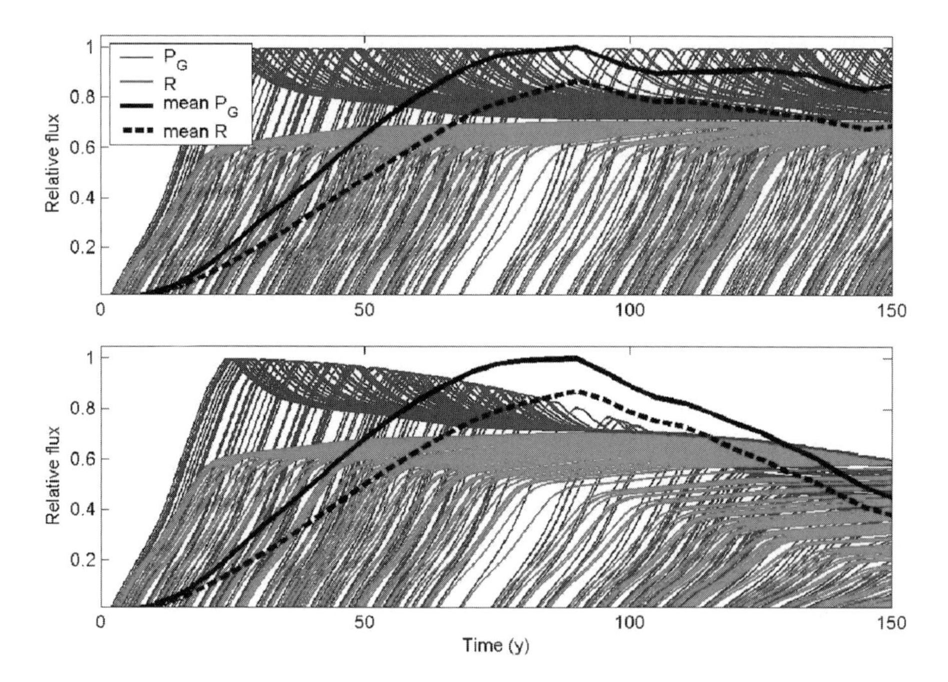

Figure 3.4 (Upper) A simulation of ecosystem C dynamics assuming an ecosystem is colonised by individuals with life histories identical to the development of ecosystems as proposed by Odum (1969). (Lower) Same as above, but assuming that the maximum growth rate of individuals progressively declines, in accordance with the assumptions of the Strategy of Ecosystem Development (Odum, 1969). Note the vigour of C uptake does not dramatically decline as ecosystems age in either case, in accordance with recent global observations (Luyssaert *et al.* 2008).

a decrease in B with age, and the SED would predict an asymptote. A recent global review of C sequestration with forest age (Luyssaert *et al.* 2008) comprehensively challenged the SED as a management tool, and clearly demonstrated that old-growth forested ecosystems sequester C on average. Ecosystem-based assumptions (the SED) do not predict this outcome, but individual-based models (e.g. Figure 3.4) do. The mechanisms responsible for the observed stability of C uptake with forest age are yet to be fully understood. These mechanisms probably relate back to the scientist's view of an ecosystem as a collection of organisms competing for resources and struggling for life, or a system that attempts to reach an equilibrium in the face of stochastic environmental and biological forcings.

Critique of ascendancy

The ascendancy concept is an excellent application of systems thinking to ecosystem ecology. The models used in its construction can be written to explicitly conserve mass and energy (thermodynamics), and ecosystem development is described in terms of information theory (statistics). The statement that ecosystems have the 'propensity' to increase in ascendancy leaves open the option that, by chance, they may not. For example, extreme events may reorganise their structure in a way that decreases information content and potential flow pathways (Gunderson and Holling 2002).

The major limitation of the ascendancy concept is well recognised by its authors (Ulanowicz 1986, Jørgensen *et al.* 2007); it is difficult if not impossible to characterise all of the flow paths required to accurately quantify ascendancy in real ecosystems. The magnitude of all pathways is rarely if ever known (Jørgensen *et al.* 2007), the existence of some pathways may not be known, and the dynamics of the unknown must be approximated. Therefore, calculations or approximations of ascendancy for a single ecosystem may differ among different investigators who may employ different box models to characterise inter-ecosystem transport.

Despite these limitations, it follows from the previous examples that older ecosystems probably contain more information and more pathways for information to be transferred. These pathways may have helped confer the observed C uptake stability at HW (Figure 3.3). These pathways logically increase with increasing forest diversity. The role of diversity in determining ecosystem stability and productivity has often been examined in ecosystems with vegetation of short stature (Tilman and Downing 1994, Hector *et al.* 1999), but rarely entire forest stands (Caspersen and Pacala 2001, Vila *et al.* 2003). Thermodynamic/statistical concepts like ascendancy are uniquely suited for interpreting why older and diverse systems may be productive and stable, but these tools have not been applied to forest ecosystem ecology to date.

Critique of the Ecological Law of Thermodynamics

The ELT is an interesting postulate, but it may overemphasise thermodynamics at the expense of statistics and thereby overestimate its ability to predict ecological behaviour. Stating the ELT as a law rather than a postulate may follow from its dependence on the proposed Fourth Law of Thermodynamics (Odum 1994), but the Fourth Law is not universally agreed upon. Ecosystems rarely obey laws that they don't have to, namely physical laws, with some authors arguing that laws are foreign to biology (Elsasser 1998). It is also difficult to incorporate the critical role of disturbance (Gunderson and Holling 2002) in the ELT framework. Whereas the ELT may explain broad patterns in nature (Jørgensen *et al.* 2007), like all theories it must be carefully considered for ecosystem applications. On the other hand, the contentions of Schneider and Kay (1994), that more mature ecosystems have lower surface temperatures, has been explored by only a few studies (Luvall and Holbo 1991, Akbari *et al.* 1999, Quattrochi and Luvall 1999, Wagendorp *et al.* 2005, Schneider and Sagan 2006) despite the clear implications for earth systems science in an era of global change.

With regard to the case study of ecological succession, the mature ecosystem (HW) probably has more exergy storage because it has the greatest basal area (Stoy *et al.* 2006a) and thereby B (noting differences in wood density between pine and oak or hickory species), but it is unclear how or if HW reached its present state by selecting the organisation that gave the highest exergy flow. HW developed from a PP-type ecosystem by stochastic dispersal, death and disturbance events. Reconciling the anthropomorphic tone of the ELT with the chance events that give rise to ecosystem structure and thereby function is not straightforward, but advanced techniques exist to incorporate stochastic events in models of ecosystem function (Porporato *et al.* 2002).

Reconciling system and individual perspectives: ecological statistical mechanics

Is there a way to incorporate thermodynamics and information theory in a way that encompasses individual effects yet retains the holistic approach for describing ecosystem behaviour? Are there statistical mechanics of ecosystems (Kerner 1957)? We know that information (information entropy) increases with ecosystem development (Ulanowicz 1986, Jørgensen *et al.* 2007). These are the 'combinations and processes offered' in the ELT. It follows from thermodynamics that local exergy increases over time in the absence of a disturbance, but it also follows from statistics that the chance that a major disturbance event occurs increases over time.

It has been argued here that individual-based stochastic models may be superior for describing ecosystem dynamics than whole-system approximations like the SED (Moorcroft *et al.* 2001, Krivov *et al.* 2003). Rather than firm ecosystem-level laws, a field of study that combines statistics (Clark 2003) and information theory (Ulanowicz 1986) with thermodynamics and physics (Jørgensen *et al.* 2007) may be able to rectify the individual and ecosystem viewpoints (Gleason 1926, Odum 1969). Let us call this combined approach to coupling statistics (information theory) and function (thermodynamics) *statistical ecological mechanics*[7] and demonstrate how modern Bayesian statistical techniques can be used to build such a framework.

As ecosystems gain mass, energy and information, more flow paths are available for the flow of mass, energy and information (Ulanowicz 1980). Individuals are the units that are struggling to survive in a Darwinian manner,

[7] Statistical ecological mechanics not statistical mechanics *sensu stricto* because ecosystems are small- or middle-number systems (Jørgensen *et al.* 2007). A critical limitation for ecological statistical mechanics as defined is that statistical mechanics was derived to explain *large-number systems*, roughly speaking those with a power greater than 10. The ability of ecological statistical mechanics to describe medium- or small-number systems will degrade, just as the macroscopic explanation for statistical mechanics degrades, as the number of elements observed decreases. In other words, statistical ecological mechanics is statistical and mechanical, but not statistical mechanical.

and individual effects must be incorporated into any model that seeks to interpret within-ecosystem energy and information flow. Assuming that structural, functional and biological diversity increases with ecosystem development, it follows that there is an increase in the *distribution* of parameters that describe an ecological function (e.g. CO_2 uptake) because information increases, following from the Shannon definition of information. This increase in information can be written:

$$V_D = V_x \sum_{i=1}^{n} p(P_i), \tag{13}$$

where $p(P)$ represents the distribution of ecosystem model parameters, V_D represents the dependent variable of interest and V_x the independent variable, noting that the independent variable may likewise have some distribution.

Upon ecosystem development, the pathway that describes the flow of highest exergy is likely to be selected because individuals that ultimately confer the ecosystem-level parameter variability (with some distribution of function) are attempting to grow and develop. This provides the Darwinian link between individual survival and growth, and ecosystem-level exergy increase (in the absence of disturbance) as a consequence. Because at its core the information content is represented by statistical models, yet the overarching model must conserve mass and energy, approaches from Hierarchical Bayes (e.g. Clark and Gelfand 2006) may be the ideal way to combine information (Figure 3.5).

Hierarchical Bayes is a logical choice for modelling information flow through ecosystems because the information entropy is defined as a distribution (e.g. Figure 3.2, Equations 9–11), and these distributions can be combined after Bayes' theorem:

$$p(A \mid B) = \frac{p(B \mid A)\, p(A)}{p(B)} \tag{14}$$

If the information inherent in an ecosystem attribute A is contingent upon other ecosystem information described by B, then the posterior probability distribution of information A given information B, written $p(A \mid B)$, is defined as the product of the prior information $p(A)$, and the probability of B given A $p(B \mid A)$ normalised by the probability of B, $p(B)$.

Thermodynamic or statistical information entropy can be formally combined at different levels of the ecosystem organisation using Equation 14. Figure 3.5 (after Clark 2005) demonstrates the general structure of an ecosystem model using these concepts from Hierarchical Bayes. Individual parameter effects are encompassed in a parameter distribution that contributes information to the parameters of a mass- and energy-conserving model. Approaches from Hierarchical Bayes for physiological and ecosystem ecology are just beginning

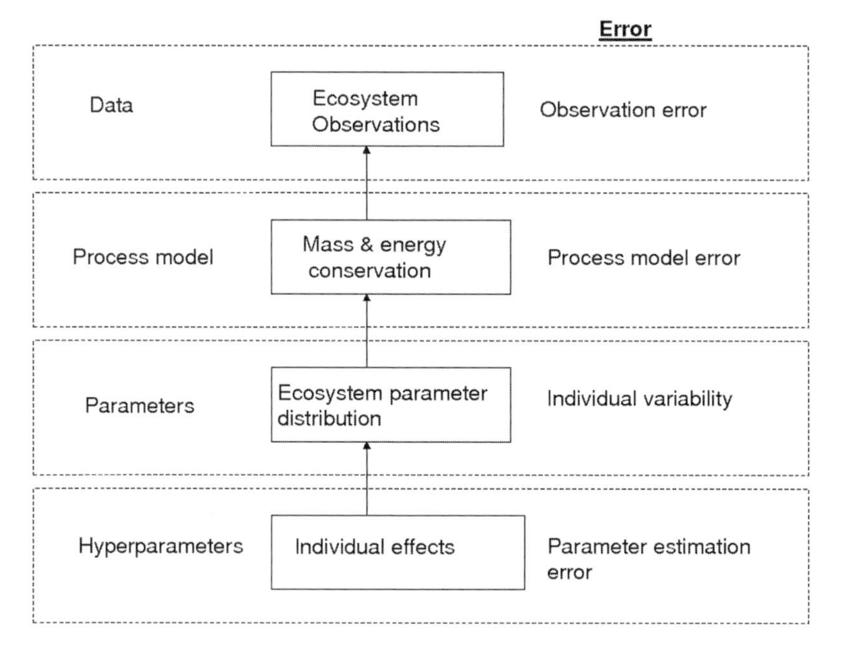

Figure 3.5 A hierarchical modelling framework for incorporating individual effects into mass- and energy-conserving models of ecosystem function.

to be explored (LaDeau and Clark 2001, Ogle and Barber 2008), but are already common in community ecology. Hierarchical Bayes is a logical and rigorous statistical technique for modelling the information and energy transfers that determine ecosystem behaviour, and will, it is hoped, find wider application in ecosystem ecology in the future.

Application: global forest management

The concepts and theories presented here have immediate applicability to global ecological science and management. Mechanisms of global policy and business move much more rapidly than ecologists' ability to study ecosystems and synthesise information (Schulze *et al.* 2000), with the result that the full ecological and social consequences of, for example, biofuel (Crutzen *et al.* 2008), are unknown until well after massive economic, social and environmental investments have been undertaken. Ecological theory provides a unifying framework for understanding the effects of industrial incursions on ecosystems to guide global policies.

Forested ecosystems have been used to demonstrate theoretical concepts here, and global forest management is at the centre of the debate over carbon sequestration for climate mitigation after the Kyoto Protocol (Conference of the Parties 3 (COP-3)) and subsequent global efforts for atmospheric CO_2 stabilisation. Signatories of the Kyoto Protocol commit to reducing CO_2 emissions or engage in emissions trading to meet emission targets. These include 'flexible mechanisms' that allow more-developed nations to undertake clean industrial projects (the Clean Development Mechanism) or other carbon projects in less-developed countries. These carbon projects include afforestation (managing forests in areas not previously forested), as agreed upon in COP-6 bis in Bonn, Germany in 2001 and COP-7 in Marrakesh, Morocco.

It has been pointed out that these 'Kyoto forests' are often comprised of young plantation stands (Schulze *et al.* 2000) which may have adverse consequences for primary forests and biodiversity (Schulze *et al.* 2002) and potentially surface temperature (Kay *et al.* 2001). A number of recent studies have questioned the ecological logic and C sequestration capacity of afforestation for C mitigation from the perspective of the coupled carbon, water and radiation budgets. The encroachment of woody vegetation into grasslands has been shown to reduce entire ecosystem C (Jackson *et al.* 2002). Forests often use more water than the vegetation that they replace; Jackson *et al.* (2005) found large net decreases in runoff from global forest plantations and evidence for large changes in soil chemistry. For example, Engel *et al.* (2005) demonstrated that an increase in water use increased soil salinity in eucalyptus plantations in Argentina. At the same time, other studies have found that reforestation of previously forested degraded lands is often beneficial. Reforestation can avoid saline groundwater incursion into upper soil layers and dry waterlogged soils to reduce flood risk (George *et al.* 1999, Plantinga and Wu 2003). Reforested agricultural fields may even lower the surface temperature (while storing more C) due to increases in evapotranspiration and water use despite concomitant increases in short-wave albedo (Juang *et al.* 2007), and recent findings demonstrate that increased canopy nitrogen content increases canopy albedo (Ollinger *et al.* 2008). The implications of afforestation/reforestation for ecosystem surface temperature have scarcely been explored despite evidence from Schneider and Kay (1994) that more complex ecosystems should have lower surface temperatures. In summary, most research demonstrates that reforestation has beneficial ecological impacts, but afforestation may not. The rate of change of ecosystem management toward afforestation that followed from global policy exceeded ecologists' ability to study the impact of these changes on ecosystems.

Recent policy developments (the 'Bali Roadmap') that resulted from the 2007 United Nations Climate Change Conference have altered the debate by suggesting that developing nations may also be given carbon credits for avoiding deforestation. The suggestion is to allow both managed (Kyoto) and unmanaged (Bali) forests to offset atmospheric C emissions.

From the perspective of the SED, the logic of the afforestation agreement and Kyoto forests is akin to shifting ecosystems to the fast growing part of the C flux/time curve (Figures 3.1 and 3.6). Avoiding deforestation through Bali forests amounts to preserving the large amounts of sequestered C in biomass (Figure 3.1), recognising that deforestation has been a major contributor to the

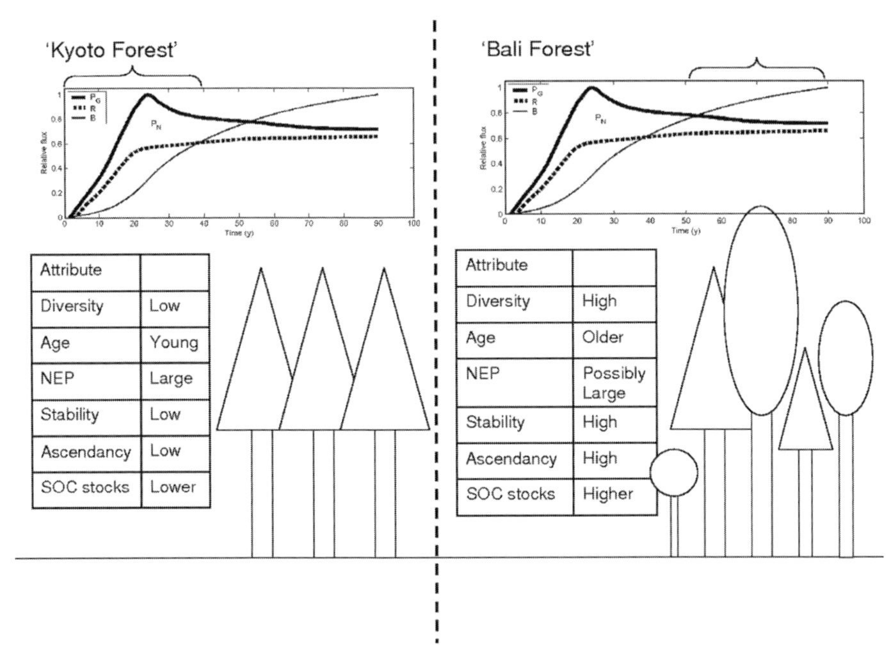

Figure 3.6 A conceptual diagram of a hypothetical 'Kyoto Forest' resulting from an afforestation project (left) and a 'Bali Forest' resulting from the preservation of mature existing forests for the additional benefit of preserving C stocks and potentially sequestration. A suite of hypothetical ecosystem attributes that follow from the Strategy of Ecosystem Development (SED (Odum 1969)) and thermodynamic concepts from ecosystem ascendancy (Ulanowicz 1986) are listed in the corresponding tables. Note that recent research debates the SED assumptions that ecosystem productivity approaches zero as ecosystems age.

increase in atmospheric CO_2 concentration above pre-industrial levels. Under these assumptions, the SED would also consider the Kyoto forests from afforestation projects to be on average more sensitive to environmental perturbations, and Bali forests less so (Figure 3.6). This may be a non-trivial distinction for actual sequestered C if many predictions of future climatic variability suggest that the likelihood of extreme events may increase (Easterling *et al.* 2000, IPCC 2007). Some authors have pointed out that biosphere C sequestration is a slow process while C losses to the atmosphere after or during disturbances such as fire, pest outbreaks, logging or extreme weather events may be relatively rapid (Gunderson and Holling 2002, Körner 2003), so the interplay between climatic sensitivity and ecosystem C sequestration may be non-trivial.

Again, it is difficult to ascertain the implications of Kyoto and Bali forests on ascendancy due to lack of data, but one may assume that the (on average) more mature and diverse Bali forests have greater ascendancy but lower developmental capacity than (on average) younger Kyoto forests (Ulanowicz and Hannon 1987, Schneider and Kay 1994). Kyoto forests may be seen as having lower interconnectivity and functional redundancy, but it is difficult to hypothesise about the practical importance of these statements because studies on the diversity and stability of vegetative communities usually focus on shorter-statured vegetation that is more amenable to experimental manipulation as discussed. Theory must fill the gap caused by the lack of studies on the role of forest diversity on productivity and resilience, and established theories must be improved and expanded to guide ecosystem management.

In summary, the ecological theories presented here are anything but esoteric; ecological theory must guide ecosystem management, and the consequences have been (and will be) rapid and global in an interconnected, changing world. The assumption made by the SED, namely that ecosystems approach 'equilibrium' with age, appears to have massively understated the importance of older forests to C sequestration and storage (Luyssaert *et al.* 2008). Rather, ecosystems are far from equilibrium systems that dissipate energy and produce entropy (Schneider and Kay 1994). The fundamentals of thermodynamics and statistics can therefore be used to improve models of ecosystem behaviour, to guide our understanding of natural systems for better ecosystem management.

References

Akbari, M., S. Murphy, J.J. Kay, and C. Swanton. 1999. Energy-based indicators of (agro) ecosystem health. In D.A. Quattrochi and J.C. Luvall (editors), *Thermal remote sensing in land surface processes*. Ann Arbor Press, Ann Arbor, Michigan.

Baldocchi, D., E. Falge, L.H. Gu, R. Olson, D. Hollinger, S. Running, P. Anthoni, C. Bernhofer, K. Davis, R. Evans, J. Fuentes, A. Goldstein, G. Katul, B. Law, X.H. Lee, Y. Malhi, T. Meyers, W. Munger, W. Oechel, K.T. Paw, U, K. Pilegaard, H.P. Schmid, R. Valentini, S. Verma, T. Vesala, K. Wilson, and S. Wofsy. 2001. FLUXNET: A new tool to study the temporal and spatial variability of ecosystem-scale carbon dioxide, water vapor, and energy flux densities. *Bulletin of the American Meteorological Society* **82**:2415–34.

Baldocchi, D.D. 2008. 'Breathing' of the terrestrial biosphere: lessons learned from a global network of carbon dioxide flux measurements systems, Turner Review. *Australian Journal of Botany* **56**:1–26.

Brillouin, L. 1953. Negentropy principle of information. *Journal of Applied Physics* **24**:1152–63.

Carey, E.V., A. Sala, R. Keane, and R.M. Callaway. 2001. Are old forests underestimated as global carbon sinks? *Global Change Biology* **7**:339–44.

Caspersen, J.P. and S.W. Pacala. 2001. Successional diversity and forest ecosystem function. *Ecological Research* **16**:895–903.

Clark, J.S. 2003. Uncertainty and variability in demography and population growth: a hierarchical approach. *Ecology* **84**:1370–81.

Clark, J.S. 2005. Why environmental scientists are becoming Bayesians. *Ecology Letters* **8**:2–14.

Clark, J.S. and A.E. Gelfand. 2006. *Hierarchical modelling for the environmental sciences*. Oxford University Press, Oxford.

Clements, F.E. 1936. Nature and structure of the climax. *Journal of Ecology* **24**:252–84.

Crutzen, P.J., A.R. Mosier, K.A. Smith, and W. Winiwarter. 2008. N_2O release from agro-biofuel production negates global warming reduction by replacing fossil fuels. *Atmospheric Chemistry and Physics* **8**:389–95.

Davidson, E.A. and I.A. Janssens. 2006. Temperature sensitivity of soil carbon decomposition and feedbacks to climate change. *Nature* **440**:165–73.

de Wit, R. 2005. Do all ecosystems maximize their distance with respect to thermodynamic equilibrium? A comment on the Ecological Law of Thermodynamics (ELT) proposed by Sven Erik Jørgensen. *Scientia Marina* **69**:427–34.

Dewar, R. 2003. Information theory explanation of the fluctuation theorem, maximum entropy production and self-organized criticality in non-equilibrium stationary states. *Journal of Physics A: Mathematical and General* **36**:631–41.

Easterling, D.R., G.A. Meehl, C. Parmesan, S.A. Cahangnon, T.R. Karl, and L.O. Mearns. 2000. Climate extremes: Observations, modeling, and impacts. *Science* **289**:2068–74.

Ekblad, A. and P. Högberg. 2001. Natural abundance of 13C reveals speed of link between tree photosynthesis and root respiration. *Oecologia* **127**:305–8.

Elsasser, W.M. 1998. *Reflections on a theory of organisms*. The Johns Hopkins University Press, Baltimore, Maryland.

Engel, V., E.G. Jobbágy, M. Stieglitz, M. Williams, and R.B. Jackson. 2005. Hydrological consequences of eucalyptus afforestation in the Argentine pampas. *Water Resources Research* **41**:W10409.

Evans, D.J. and D.J. Searles. 1994. Equilibrium microstates which generate second law violating steady states. *Physical Review E* **50**:1645–8.

George, R.J., R.A. Nulsen, R. Ferdowsian, and G.P. Raper. 1999. Interactions between trees and groundwaters in recharge and discharge areas – A survey of Western Australian sites. *Agricultural Water Management* **39**:91–113.

Gleason, H.A. 1926. The individualistic concept of the plant association. *Bulletin of the Torrey Botanical Club* **53**:7–26.

Gunderson, L.H. and C.S. Holling, (eds). 2002. *Panarchy: understanding transformations in human and natural systems*. Island Press, Washington DC.

Hector, A., B. Schmid, C. Beierkuhnlein, M.C. Caldeira, M. Diemer, P.G. Dimitrakopoulos, J.A. Finn, H. Freitas, P.S. Giller, J. Good, R. Harris, P. Högberg, K. Huss-Danell, J. Joshi, A. Jumpponen, C. Korner, P.W. Leadley, M. Loreau, A. Minns, C.P.H. Mulder, G. O'Donovan, S.J. Otway, J.S. Pereira, A. Prinz, D.J. Read, M. Scherer-Lorenzen, E.D. Schulze, A.S.D. Siamantziouras, E.M. Spehn, A.C. Terry, A.Y. Troumbis, F.I. Woodward, S. Yachi, and J.H. Lawton. 1999. Plant diversity and productivity experiments in European grasslands. *Science* **286**:1123–7.

Hedin, L.O., J.C. von Fischer, N.E. Ostrom, B.P. Kennedy, M.G. Brown, and G.P. Robertson. 1998. Thermodynamic constraints on nitrogen transformations and other biogeochemical processes at soil–stream interfaces. *Ecology* **79**:684–703.

Högberg, P., A. Nordgren, N. Buchmann, A.F.S. Taylor, A. Ekblad, M.N. Högberg, G. Nyberg, M. Ottoson-Löfvenius, and D.J. Read. 2001. Large-scale forest girdling shows that current photosynthesis drives soil respiration. *Nature* **411**:789–92.

IPCC. 2007. *Climate change 2007 – the physical science basis: working group I contribution to the fourth assessment report of the IPCC*. Cambridge University Press.

Jackson, R., E.G. Jobbágy, R. Avissar, S. Baidya Roy, D. Barrett, C.W. Cook, K.A. Farley, D.C. le Maitre, B.A. McCarl, and B.C. Murray, 2005. Trading water for carbon with biological carbon sequestration. *Science* **310**:1944–7.

Jackson, R.B., J.L. Banner, E.G. Jobbágy, W.T. Pockman, and D.H. Wall. 2002. Ecosystem carbon loss with woody plant invasion of grasslands. *Nature* **418**:623–6.

Jaynes, E.T. 1957. Information theory and statistical mechanics. *Physical Review* **106**:620–30.

Jaynes, E.T. 2003. *Probability theory: the logic of science*. Cambridge University Press.

Jørgensen, S.E. 1997. *Integration of ecosystem theories: a pattern*. 2nd edition. Kluwer Academic Publishers, Amsterdam, The Netherlands.

Jørgensen, S.E., J.C. Marques, F. Müller, S.N. Nielsen, P.C. Patten, E. Tiezzi, and R.E. Ulanowicz. 2007. *A new ecology: systems perspective*. p. 275. Elsevier, Amsterdam, The Netherlands.

Juang, J.-Y., G.G. Katul, M.B.S. Siqueira, P.C. Stoy, and K.A. Novick. 2007. Separating the effects of albedo from eco-physiological changes on surface temperature along a successional chronosequence in the southeastern US. *Geophysical Research Letters* **34**:doi:10.1029/2007GL031296.

Kay, J.J., T.F.H. Allen, R. Fraser, J.C. Luvall, and R.E. Ulanowicz. 2001. Can we use energy based indicators to characterize and measure the status of ecosystems, human, disturbed and natural? pp. 121–33 in *Proceedings of the international workshop: Advances in energy studies: exploring supplies, constraints and strategies*, Porto Venere, Italy, 23–27 May, 2000.

Kerner, E. 1957. A statistical mechanics of interacting biological species. *Bulletin of Mathematical Biophysics* **19**:121–46.

Kira, T. and T. Shidei. 1967. Primary production and turnover of organic matter in different forest ecosystems of the western Pacific. *Japanese Journal of Ecology* **17**:70–87.

Kleidon, A. In press. Global energy balance. In S.E. Jørgensen (editor), *Encyclopedia of Ecology*. Elsevier, Amsterdam, The Netherlands.

Knohl, A., E.-D. Schulze, O. Kolle, and N. Buchmann. 2003. Large carbon uptake by an unmanaged 250-year-old deciduous forest in Central Germany. *Agricultural and Forest Meteorology* **118**:151–67.

Körner, C. 2003. Slow in, rapid out – carbon flux studies and Kyoto targets. *Science* **300**:1242–3.

Krivov, S., R.E. Ulanowicz, and A. Dahiya. 2003. Quantitative measures of organization for multiagent systems. *Biosystems* **69**:39–54.

LaDeau, S. and J.S. Clark. 2001. Rising CO_2 levels and the fecundity of forest trees. *Science* **292**:95–8.

Lehinger, A. 1971. *Bioenergetics – the molecular basis for biological energy transformations*. The Benjamin/Cummings Publishing Company, London.

Lotka, A.J. 1922a. Contribution to the energetics of evolution. *Proceedings of the National Academy of Sciences of the United States of America* **8**:147–51.

Lotka, A.J. 1922b. Natural selection as a physical principle. *Proceedings of the National Academy of Sciences of the United States of America* **8**:151–4.

Luvall, J.C. and H.R. Holbo. 1991. Thermal remote sensing methods in landscape ecology. In M.G. Turner and R.H. Gardner (editors), *Quantitative methods in landscape ecology*. Springer-Verlag Heidelberg.

Luyssaert, S., E.D. Schulze, A. Borner, A. Knohl, D. Hessenmoller, B.E. Law, P. Ciais, and J. Grace. 2008. Old-growth forests as global carbon sinks. *Nature* **455**:213–15.

Magnani, F., M. Mencuccini, and J. Grace. 2000. Age-related decline in stand productivity: the role of structural acclimation under hydraulic constraints. *Plant Cell and Environment* **23**:251–63.

McCarthy, H.R., R. Oren, K.H. Johnsen, S.G. Pritchard, M.A. Davis, C. Maier, and H.-S. Kim. 2006. Ice storms and management practices interact to affect current carbon sequestration in forests with potential mitigation under future CO_2 atmosphere.

Journal of Geophysical Research-Atmospheres **111**:doi:10.1029/2005JD006428.

Moorcroft, P.R. 2003. Recent advances in ecosystem–atmosphere interactions: an ecological perspective. *Proceedings of the Royal Society of London Series B-Biological Sciences* **270**:1215–27.

Moorcroft, P.R., G.C. Hurtt, and S.W. Pacala. 2001. A method for scaling vegetation dynamics: the ecosystem demography model (ED). *Ecological Monographs* **71**:557–86.

Naiman, R.J. and H. Décamps. 1997. The ecology of interfaces: riparian zones. *Annual Review of Ecology and Systematics* **28**:621–58.

Odum, E.P. 1969. The strategy of ecosystem development. *Science* **164**:262–70.

Odum, E.P. 1971. *Fundamentals of Ecology*. W.B. Saunders, Philadelphia.

Odum, H.T. 1994. *Ecological and general systems: an introduction to systems ecology*. Colorado University Press, Niwot, Colorado.

Ogle, K. and J.J. Barber. 2008. Bayesian data-model integration in plant physiological and ecosystem ecology. In U. Lüttge, W. Beyschlag, and J. Murata (editors), *Progress in Botany*. Springer, Berlin, Heidelberg.

Ollinger, S.V., A.D. Richardson, M.E. Martin, D.Y. Hollinger, S. Frolking, P.B. Reich, L.C. Plourde, G.G. Katul, J.W. Munger, R. Oren, M.-L. Smith, K.T. Paw U, P.V. Bolstad, B.D. Cook, M.C. Day, T.A. Martin, R.K. Monson, and H.P. Schmid. 2008. Canopy nitrogen, carbon assimilation, and albedo in temperate and boreal forests: functional relations and potential climate feedbacks. *Proceedings of the National Academy of Sciences of the United States of America* **105**: 19336–41.

Plantinga, A.J. and J. Wu. 2003. Co-benefits from carbon sequestration in forests: evaluating reductions in agricultural externalities from an afforestation policy in Wisconsin. *Land Economics* **79**:74–85.

Porporato, A., P. D'Odorico, F. Laio, L. Ridolfi, and I. Rodriguez-Iturbe. 2002. Ecohydrology of water-controlled ecosystems. *Advances in Water Resources* **25**:1335–48.

Prigogine, I. 1961. *Introduction to thermodynamics of irreversible processes*. 2nd edition. Interscience Publishers New York.

Quattrochi, D.A. and J.C. Luvall. 1999. Thermal infrared remote sensing for analysis of landscape ecological processes: methods and applications. *Landscape Ecology* **14**:577–98.

Reichstein, M., D. Papale, R. Valentini, M. Aubinet, C. Bernhofer, A. Knohl, T. Laurila, A. Lindroth, E. Moors, K. Pilegaard, and G. Seufert. 2007. Determinants of terrestrial ecosystem carbon balance inferred from European eddy covariance flux sites. *Geophysical Research Letters* **34**:doi:10.1029/2006GL027780.

Röser, C., L. Montagnani, E.D. Schulze, D. Mollicone, O. Kolle, M. Meroni, D. Papale, L.B. Marchesini, S. Federici, and R. Valentini. 2002. Net CO_2 exchange rates in three different successional stages of the 'Dark Taiga' of central Siberia. *Tellus B* **54**:642–54.

Schneider, E.D. and J.J. Kay. 1994. Life as a manifestation of the second law of thermodynamics. *Mathematical and Computer Modelling* **19**:25–48.

Schneider, E.D. and D. Sagan. 2006. *Into the cool: energy flow, thermodynamics, and life*. University of Chicago Press, Chicago.

Schrödinger, E. 1944. *What is life? – The physical aspect of the living cell*. Cambridge University Press.

Schulze, E.-D., C. Wirth, and M. Heimann. 2000. Managing forests after Kyoto. *Science* **289**:2058–9.

Schulze, E.D., R. Valentini, and M.J. Sanz. 2002. The long way from Kyoto to Marrakesh: Implications of the Kyoto Protocol negotiations for global ecology. *Global Change Biology* **8**:505–18.

Shannon, C.E. 1948. A mathematical theory of communication. *Bell System Technical Journal* **27**:379–423 and 623–56.

Stoy, P.C., G.G. Katul, M.B.S. Siqueira, J.-Y. Juang, H.R. McCarthy, A.C. Oishi, J.M. Uebelherr, H.-S. Kim, and R. Oren. 2006a. Separating the effects of climate and vegetation on evapotranspiration along a successional chronosequence in the southeastern U.S. *Global Change Biology* **12**:2115–35.

Stoy, P.C., G.G. Katul, M.B.S. Siqueira, J.-Y. Juang, K. Novick, H.R. McCarthy, A.C. Oishi, and R. Oren. 2008. The role of vegetation in determining carbon sequestration along ecological succession in the southeastern United States. *Global Change Biology* **14**:1409–27.

Stoy, P.C., G.G. Katul, M.B.S. Siqueira, J.-Y. Juang, K.A. Novick, and R. Oren. 2006b. An evaluation of methods for partitioning eddy covariance-measured net ecosystem exchange into photosynthesis and respiration. *Agricultural and Forest Meteorology* **141**:2–18.

Tilman, D. and J.A. Downing. 1994. Biodiversity and stability in grasslands. *Nature* **367**:363–5.

Ulanowicz, R.E. 1980. An hypothesis on the development of natural communities. *Journal of Theoretical Biology* **85**:223–45.

Ulanowicz, R.E. 1986. *Growth & Development: Ecosystems Phenomenology*. Springer Verlag, New York.

Ulanowicz, R.E., S.J. Goerner, B. Lietaer, and R. Gomez. In press. Quantifying sustainability: resilience, efficiency and the return of information theory. *Ecological Complexity* **6**: 27–36.

Ulanowicz, R.E. and B.M. Hannon. 1987. Life and the production of entropy. *Proceedings of the Royal Society of London Series B, Biological Sciences* **232**:181–92.

Vilá, M., J. Vayreda, C. Gracia, and J.J. Ibáñez. 2003. Does tree diversity increase wood production in pine forests? *Oecologia* **135**:299–303.

Wagendorp, T., H. Gulinck, P. Coppin, and B. Muys. 2005. Land use impact evaluation in life cycle assessment based on ecosystem thermodynamics. *Energy* **31**:112–25.

Wang, G.M., E.M. Sevick, E. Mittag, D.J. Searles, and D.J. Evans. 2002. Experimental demonstration of violations of the second law of thermodynamics for small systems and short time scales. *Physical Review Letters* **89**:050601.

Zhou, G., S. Liu, Z. Li, D. Shang, X. Tang, C. Zhou, J. Yan, and J. Mo. 2006. Old-growth forests can accumulate carbon in soils. *Science* **314**:1417.

CHAPTER FOUR

Ecosystem health

PIRAN C. L. WHITE, JAMES C. R. SMART,
and DAVID G. RAFFAELLI
Environment, University of York
ANNA R. RENWICK
College of Life Science and Medicine, University of Aberdeen

Introduction: origins and development of the ecosystem health concept

The need to understand and quantify ecosystem behaviour and condition has come to the forefront of environmental policy due to a greater emphasis on environmental sustainability and an accompanying recognition of the scarcity of natural resources, such as water, soil and biological diversity. Increasing concern regarding human impacts on the environment and the possibility that some human-induced changes in ecological systems may be irreversible has also focused attention on ways in which such changes can be assessed and, if possible, avoided. From the policy maker's perspective, the concern is not only in terms of the possible extent of these problems, but also the likelihood of their occurrence and the timeframe over which they may operate. In the context of global climate change, understanding the functioning of ecosystems, and how their health and performance can be measured and monitored over time are of critical importance, since these are linked inextricably with human health and well-being.

A considerable body of literature over the past decade has sought to define ecosystem health in practical terms. The majority of the definitions of ecosystem health concentrates exclusively on ecological aspects. For example, Costanza (1992) defines the term as follows: 'An ecological system is healthy and free from "distress syndrome" if it is stable and sustainable - that is, if it is active and maintains its organisation and autonomy over time and is resilient to stress.' However, increased recognition of the interdependence of human and natural systems has provided the impetus for a broader definition, encompassing the biological, economic and human dimensions of the system. Thus, Xu and Mage (2001) included each of these dimensions in their definition of the health of managed systems: 'the system's ability to realise its functions desired by society and to maintain its structure needed both by its functions and by society over a long time'. This definition considers both functional (activities and processes, which occur within an ecosystem, for example gross productivity) and structural characteristics (individual components of the system and their interrelationship,

Ecosystem Ecology: A New Synthesis, eds. David G. Raffaelli and Christopher L. J. Frid. Published by Cambridge University Press.

for example species diversity) of the ecosystem in the context of societal needs as well as emphasising the importance of temporal changes (ibid.).

Approaches for assessing ecosystem health

A number of different approaches have been taken to assess ecosystem health, two of which are discussed below. Costanza (1992) proposed that a full assessment of the 'health' of an ecosystem would account for the following six attributes of the system concerned: (1) homeostasis (self-regulation); (2) absence of disease; (3) diversity or complexity (number and types of species); (4) stability or resilience; (5) vigour or scope for system growth; and (6) balance between system components. He considered it necessary to address all, or at least a majority, of these attributes simultaneously and proposed an index (*HI*) which reflected the ability of a healthy and sustainable system to maintain its metabolic activity level ('system vigour' *V*), as well as its internal structure and organisation ('system organisation' *O*), and its resilience to outside stresses ('system resilience' *R*) over the spatial and temporal frames of reference. Thus:

$$HI = V \cdot O \cdot R$$

Application of Costanza's ecosystem health index requires the assessment of vigour, structure and resilience in a quantified and commensurable fashion in real ecosystems. Each individual component poses different challenges for quantification. Costanza suggested that it might be appropriate to use different quantities as measures of vigour, organisation and resilience within different ecosystem settings (Table 4.1) depending on the data available and the perspective or value system adopted for the health assessment. However, such flexibility could impede ready comparison of ecosystem health indices between ecosystem settings and could also introduce weighting difficulties between the vigour, organisation and resilience components within the health index in any particular setting.

A related approach advocated by Xu and Mage (2001) used four sets of general criteria to assess the health of managed ecosystems: structural, functional, organisational (condition of ecosystem relative to its relationships with the external environment), and dynamics (the temporal aspect of the ecosystem and its ability to cope with change) (Table 4.2). Structural and functional changes within a stressed ecosystem will induce corresponding changes in system level attributes, such as ascendancy, buffer capacities, exergy and structural exergy, terms which may be unfamiliar to many readers and which are defined later in the chapter (Jørgensen 1997).

Indicators and indices derived from direct measurement

Ecological indicators

Ecological indicators are commonly used to determine the impact of various environmental contaminants and disturbances, and such indicators have been

Table 4.1. *Vigour (V), organisation (O) and resilience (R) components proposed for use in the construction of an Ecosystem Health Index (HI), in a variety of settings (based on Table 2 in Costanza 1992).*

Component	Related concepts	Potential measures	Field of origin	Assessment method
Vigour	Function	GPP, NPP, GEP	Ecology	Direct measurement
	Productivity	GNP, NNP*	Economics	Estimation from data
	Throughput	Metabolism	Biology	system modelling
Organisation	Structure	Diversity index	Ecology	Network analysis
	Biodiversity	Connectance		
Resilience		Proof against change of state	Ecology	Simulation modelling
			Economics	
		Return time		

* Gross Primary Production, Net Primary Production, Gross Ecosystem Production, Gross National Product, Net National Product respectively.

applied to assess the overall health of an ecosystem (Xu *et al.* 1999). Several different types of ecological indicator exist and those chosen depend on their required role in the assessment process. Ecological indicators can be divided into three classes: (1) *early warning indicators* that detect impending changes; (2) *compliance indicators* that detect changes in characteristics beyond acceptable limits; and (3) *diagnostic indicators* that show the causes of deviations (Boulton 1999). An ideal indicator should incorporate the following characteristics to provide a holistic interpretation of the status of the monitored system (ibid.): (1) relevant, robust and scientifically supported (Walker 2002); (2) readily repeatable and easily validated; (3) relatively cheap and quick to measure; (4) amenable to measurement by non-trained persons; and (5) able to inform ecosystem managers and policy makers about the state of the ecosystem (Bockstaller and Girardin 2003). An indicator should also be unambiguous in its response to threats to the system, although this may not always be possible given the complexity of the systems being monitored (Sueter 1993). Indicators should use standard units (van der Werf and Petit 2002) and be tested and calibrated against empirical measures to determine their validity (Duelli and Obrist 2003). This validation process is important, but it is rarely implemented (Bockstaller and Girardin 2003). Surrogate indicators are often used to simplify data collection and case studies of well-known ecosystems may be used to evaluate the utility of such simplifications (van der Werf and Petit 2002).

The usefulness of ecological indicators depends on the approach and spatial scale adopted as well as their practicality in use. Ecosystems can potentially exist in multiple dynamic states (Patil *et al.* 2002), and ecological indicators

Table 4.2. *General criteria for assessing ecosystem health (Xu and Mage 2001 with some amendments).*

Criterion	Related concepts	Definition	Relation to health	Examples of measures
Structural	Resource availability	Volume of resources necessary to achieve or maintain system functions	Higher level of resource availability considered healthier	Crop production
	Resource accessibility	Ease of access to and utilisation of system's resources	High accessibility to all resources considered healthier	Accessibility of agricultural land to water (Ali 1995)
	Diversity	Number of system components and how they vary across space	More diverse ecosystem is healthier	Number of species and size of population
	Equality*	Evenness of the distribution of ecosystem products across society (evenness may be a measure of diversity)	More evenness or equality is healthier	Quantified assessment of land resource distribution among households (Conway 1985)
	Equity	Distributive fairness of resources among people across space and time	More equitable system is healthier	Societal perception of fairness of distribution of resources
Functional	Productivity	Output of product per unit of resource input	Higher productivity is healthier	Energy output per hectare
	Efficiency	Ratio of output (product) to input (cost)	Highly efficient system is healthier	Energy ratio, yield, net income per unit of resource (Conway 1987)
	Effectiveness	Capacity of the system to meet goals of stakeholder	More effective system is healthier	Quantified assessment of the outcome of implemented programmes or conservation schemes

Organisational	Integrity	No commonly accepted definition		
	Self-organisation	Degree to which a system maintains its organisation	Largely conceptual, no practical measure	
	Autonomy	Ability of system to absorb external disturbance by self-reorganising its structural and functional units and continue to function by self-regulating the flow of energy, information and materials	Similar to self-organisation	Quantified measures of the state of rural communities despite agricultural intensification in the area
	Self-dependence/self-reliance	Relationship of system's organisation with the external environment	Greater self-dependence is healthier	Fossil fuel input per hectare
Dynamics	Stability	Ability of a system to attain or retain an equilibrium steady state (Holling 1986)	More stable is healthier	Measure of the change in socio-economic and/or biophysical environments over time
	Resilience	Ability of an ecosystem to cope with natural and socio-economic stresses (Waltner-Toews 1994)	More resilient system is healthier; few practical or numerical measures	
	Capacity to respond	Capacity of a system with respond to various stresses	Essentially synonymous with resilience	

* Defined as Equitability by Xu and Mage (2001), although equitability is actually the same as equity in this context.

are therefore required to provide meaningful assessments in the face of such changes. Time series data are often required and spatial scales ranging from habitat patches to landscape, regional and global scales need to be considered (Waldhardt *et al.* 2003) in order to address issues at the ecosystem, regional, national and international levels.

Multimetric indices

Purely biological assessments which use a single or limited number of species are insufficient to capture the complexity of living systems (Buchs 2003) and human dependence on them (Haskell *et al.* 1992, Rapport *et al.* 1999). More holistic approaches using sets of indicators that incorporate economic, ecological and societal components have therefore been devised (Stork 1995, Stork and Eggleton 1992, Ferris and Humphrey 1999). Each individual indicator is selected to represent a different aspect of ecosystem health, and the simultaneous use of several indicators provides a better measurement of the overall health of the ecosystem which encompasses both biophysical and socio-economic aspects (Jørgensen 1997, Rapport 1995, Karr 1992). The values of individual indicators can then be amalgamated to produce a multimetric index (Duelli and Obrist 2003), e.g. gross ecosystem product (GEP) (Hannon 1985); ecosystem stress indicators (Rapport *et al.* 1985); the index of biotic integrity (IBI) (Karr *et al.* 1986); Costanza's overall index of ecosystem health (HI) (Costanza 1992); buffer capacities (Jørgensen 1995, Jørgensen *et al.* 1995).

Jørgensen (1997) proposed that a number of these composite indices could be analysed simultaneously to obtain a full picture of ecosystem health. However, combining separate indicators into a multimetric index must be done with care, particularly if the separate indicators do not track each other with respect to health. Hoffmann and Greef (2003) and Hoffmann *et al.* (2003) developed a mosaic indicator approach, based on qualitative and quantitative assessments, that acknowledged the historical development of the landscape. In a different approach, Bockstaller *et al.* (1997) proposed the use of sustainability polygons, webs or radars to overcome these issues and to aid in visual presentation of outputs (Swete-Kelly 1996, Bockstaller *et al.* 1997, Gomez *et al.* 1996). These representations show the scores of each index component simultaneously and thus prevent the need to aggregate scores across different scales (Rigby *et al.* 2001). The resultant picture integrates all the behaviour and processes of the separate elements within the biological system (Karr 1999).

Multivariate statistical approaches have also been used to capture patterns of change in the assemblages of species present under different disturbance levels. For example, aquatic invertebrate assemblages can be sampled from streams and rivers and the relative abundance of the species present compared with that expected under undisturbed conditions, deviations from the expected implying a disturbed system. Several such packages have been developed, for example, RIVPACS (Wright 1995), AusRivAS (Parsons and Norris 1996), and BEAST (Reynoldson *et al.* 1995) and they can be used as multimetric indices of the status of at least part of the ecosystem.

Problems with the use of biological monitoring and indicators

The very complexity of ecosystems makes the assessment of ecosystem health challenging. Different systems respond uniquely to stress and have certain features that are vital for their individual integrity (Rapport *et al.* 1999). Previous attempts to monitor ecosystem health have experienced a number of problems that limit their usefulness: determining which features characterise a healthy ecosystem (Belaoussoff and Kevan 1998); absence of important data (knowledge gaps); restriction of studies to small areas (Wichert and Rapport 1998); natural fluctuations in the system (Buchs 2003); determination of baseline reference points (ibid.); lack of appropriate analytical methods (Patil and Myers 1999, Patil *et al.* 2002); and the failure to integrate human, social and ecological dimensions (Epstein and Rapport 1996, Huq and Colwell 1996). The conditions within a site are essentially unique and thus results obtained for a particular site cannot be extrapolated across locations.

In order to quantify the magnitude and direction of change within the system it may be possible to relate results to a presumed control site (Holland *et al.* 1994, Steinmann and Gerowitt 2000, Bartels and Kampmann 1994) but suitable control sites are often difficult to find. Long-term monitoring programmes may be required to detect environmental change but this is labour intensive and expensive (Piorr 2003). Financial restrictions dictate that a compromise must generally be drawn between the precision and accuracy delivered by a particular indicator or monitoring technique, the implementation time required and its ease of use by non-specialist personnel (Buchs 2003).

Using models to assess ecosystem health

Indices and indicators of ecosystem structure and function are typically derived by direct measurement, followed in some cases by appropriate calculation. System-level metrics which have been used to assess ecosystem health are, however, generally derived from models of the ecosystem concerned. Such models should embody key ecosystem components, reflect their interrelationships appropriately and be calibrated using data from relevant study sites. In addition, model predictions should be verified by supporting measurements. Once appropriate calibrated models have been constructed, then the desired system-level metrics can be determined. Jørgensen (1997) suggested that structural, functional and system-level metrics should be applied simultaneously. This means that direct measurements and appropriate ecological modelling should be undertaken together to produce a reliable assessment of ecosystem health on the basis of Costanza's six attributes (Costanza 1992). Advances in statistical and computational methods which allow both spatial and temporal aspects of indicators to be represented may facilitate the success of ecosystem health assessments (Johnson *et al.* 2002, Patil 2000, Patil 2001a, Patil 2001b, Patil 2001c).

One group of models that has potential for ecosystem health assessment is that associated with systems ecology. Such models represent the trophic networks that connect different species in a system. Higher-level properties can be calculated from the complexity of the network and the magnitude of flows of material or information through the network (Stoy, this volume, Jørgensen *et al.* 2007), and these properties can be related to aspects of ecosystem health. Much of the terminology and concepts stem from the work of Eugene Odum (Odum 1953 – see also Odum 1969, 1985), often described as the 'father of modern ecosystem ecology'. His conjectures concerning the ways in which ecosystems move away from their thermodynamic equilibrium as they develop and 'mature' underpin much of the modelling approach to ecosystem health.

Odum's conjectures and mass-balance approaches

Over the past fifty years, ecosystem ecologists have described a range of system attributes that may have potential as indicators of ecosystem health. These include many of the original twenty-four conjectures of Odum (1969) and developments thereof, such as network ascendancy (which can be thought of as the product of the amount of material flowing through a system and that system's complexity) (Ulanowicz 1986, 1992) and exergy (which can be thought of as the 'useful' energy which must be dissipated in order to sustain an ecosystem) (Jørgensen 1995, Nielsen and Ulanowicz 2000). Central to these concepts is the view that ecosystems move progressively through developmental stages towards their mature state, culminating in a stable system with maximum biomass and/or 'information' and optimal utilisation of available energy through internalisation of material flows and increased feedback control as the system matures (Table 4.3). Analogues of many of these measurements can be determined in a straightforward manner from a mass-balance or a network model of the ecosystem under investigation (Christensen 1995, Christensen and Walters 2004). However, world views of stability and ecosystem development exist which differ from those of Odum, and these are briefly described in the next section (see also Raffaelli and Frid, this volume).

Adaptive cycles and resilience

A somewhat different perspective of ecosystem development and behaviour to that of traditional systems theory comes from recent developments in the fields of resilience and adaptive cycles (Holling 1973, Holling 1992, Gunderson and Pritchard 2002, Gunderson and Holling 2002, Raffaelli and Frid this volume). Here, ecosystems are characterised by an adaptive cycle of change that has four main phases: exploitation, conservation, release and reorganisation. Two of these phases, *exploitation* and *conservation* are characterised by species with r

Table 4.3. *Odum's twenty-four attributes of development through ecological succession (Odum 1969).*

	Ecosystem attributes	Early stages	Mature stages
Community energetics			
1.	Gross production / community respiration (P/R ratio)	Greater or less than 1	Approaches 1
2.	Gross production / standing crop biomass (P/B ratio)	High	Low
3.	Biomass supported / unit energy flow	Low	High
4.	Net community production (yield)	High	Low
5.	Food chains	Linear, mostly grazing	Web-like, predominantly detritus
Community structure			
6.	Total organic matter	Small	Large
7.	Inorganic nutrients	Extrabiotic	Intrabiotic
8.	Species diversity – variety component	Low	High
9.	Species diversity – equitability component	Low	High
10.	Biochemical diversity	Low	High
11.	Stratification and spatial heterogeneity (pattern diversity)	Poorly organised	Well organised
Life history			
12.	Niche specialisation	Broad	Narrow
13.	Size of organism	Small	Large
14.	Life cycles	Short, simple	Long, complex
Nutrient cycling			
15.	Mineral cycles	Open	Closed
16.	Nutrient exchange rate between organisms and the environment	Rapid	Slow
17.	Role of detritus in nutrient regeneration	Unimportant	Important
Selection pressure			
18.	Growth form	For rapid growth ('r-selection', e.g. weeds)	For feedback control ('K-selection', e.g. trees)
19.	Production	Quantity	Quality
Overall homeostasis			
20.	Internal symbiosis	Undeveloped	Developed
21.	Nutrient conservation	Poor	Good
22.	Stability (resistance to external perturbations)	Poor	Good
23.	Entropy	High	Low
24.	Information	Low	High

(high growth rates, high fecundity, short-lived, high dispersal, low competitive ability – 'weeds') and K (slow growth, lower fecundity, high competitive ability, long-lived – 'trees') strategies, respectively. These two phases have clear parallels with Odum's (1969) ideas about the life forms that characterise early and late developmental stages (ibid. Table 4.3). However, a unique feature of the adaptive cycle is that external and/or intrinsic perturbations (e.g. hurricanes, drought, pests and disease outbreaks) cause a sudden and catastrophic release of the accumulated biomass and materials in the system, the *release* phase (for example, a fully developed and mature forest may collapse). This material is then *reorganised* as it becomes opportunistically captured by pioneer species. In a sustainable system, the resources accumulated during the conservation phase, which determine the ecological potential of the system, may generate a similar ecosystem following reorganisation. However, if specific accumulated resources (e.g. key taxa or soil conditions) are dramatically changed during the release phase, then the system is expected to develop quite differently following reorganisation: the system would 'flip' into a different state.

Resilience of the system (its ability to cope with perturbations) will vary throughout the adaptive cycle as the system moves between the four phases. Thus, resilience is high in the reorganisation and exploitation phases, but declines during the conservation phase with the system becoming more vulnerable to 'surprises' (fire, drought, disease) as it becomes more rigid and inflexible.

This view of ecosystem behaviour is proving attractive to many ecologists and economists, partly because of the parallels and analogies which can be drawn with the behaviour of economic and social systems, thereby providing an opportunity for integrating human and ecological systems (e.g. Gunderson and Holling 2002), and partly because there is mounting evidence of such cyclic behaviour and alternate states in real ecological systems (reviewed in Gunderson and Holling 2002, Gunderson and Pritchard 2002 and references therein). Whilst adaptive cycles provide a powerful conceptual framework in which to think about how best to approach sustainable management of ecological and socio-economic systems (e.g. Gallopin 2002, Holling *et al.* 2002), we are a long way from developing operational measures of ecosystem health based on this framework.

System-level metrics from mass-balance models

Mass-balance models are relevant to ecosystem health because they provide a quantified description of the structure and function of ecosystems in their steady state and a starting point for the assessment of their health, and they can be used to explore the mechanisms which underpin ecosystem growth and development. An implicit link exists between an ecosystem's ability to follow its 'normal' path of growth and development (Table 4.3) and its 'state of health'.

Thus, Ulanowicz (1992) defines a healthy ecosystem as 'one whose trajectory toward the climax community is relatively unimpeded and whose configuration is homeostatic to influences that would displace it back to earlier successional stages'.

The construction of mass-balance models is facilitated by the readily available software package Ecopath (Fisheries Centre, University of British Columbia 2004), although other network software is also available, e.g. NET-WRK 4.2a (Ulanowicz and Kay 1991). Marine fisheries provided the initial application area for Ecopath models, and marine implementations still comprise the majority of installations (see *Ecological Modelling* Special Issue 172, 2004), although Ecopath models have now also been published for freshwater and terrestrial ecosystems (Ruesink *et al.* 2002, Dalsgaard *et al.* 1995, Christensen 1995).

The Ecopath approach to mass balance seeks to establish a balance in biomass production and consumption between user-defined groups (age-classes within a species, individual species or groups of species) in the trophic structure of the ecosystem, and also to establish a balance in energy flow within each group. The model thus requires basic estimates of the biomass, production and consumption of all groups. Some parameters can/must be estimated by the software itself, and there are routines that allow an assessment of parameter uncertainty. A full account of Ecopath and mathematical descriptions of the ecosystem metrics which it produces can be found at www.ecopath.org.

The approach is data-intensive for a large complex system so that simplified or aggregated groupings tend to be employed rather than individual species (although changes in taxonomic resolution of groupings may significantly affect the value of some of the higher-level metrics that are derived (Abarca-Arenas and Ulanowicz 2002)). Parameters can be derived by empirical investigation or by reference to other databases or established relationships, such as that between body size and production per unit biomass (P/B ratio).

Various metrics can be derived from mass-balance models that might reflect the model ecosystem's health in terms of its vigour (productivity), organisation (structure) and resilience and which are analogous to Odum's conjectures (Christensen 1995). However, two metrics, ascendancy (Ulanowicz 1986, 1980) and exergy (Jørgensen 1986, 1988a, 1988b, 1992) have been the most explored. In common with all modelling approaches the output must always be evaluated in terms of the quality of the input data and the degree of assumptions/abstractions that had to be made to construct the model. Unfortunately many applications do not provide this information and differences in basic data quality and the assumptions made make comparisons between Ecopath models of different ecosystems problematic (Robinson and Frid 2003). However, there is more scope for comparisons in the dynamics (i.e. health) of a single modelled system over time and this type of comparison will meet many of the needs of policy.

Exergy expresses the 'useful' energy which must be dissipated to sustain an ecosystem, and which is embodied within the biomass resident within that ecosystem. Ascendancy is a measure of the flows through and between the biomass compartments scaled by the system's complexity. Biomass within compartments will accumulate or diminish as a consequence of net energy flow into and out of compartments, and, conversely, the organisation inherent within the biomass within any compartment will constrain energy or nutrient flow through that compartment. Thus, exergy and ascendancy are both related to the organisational structure of biomass within, and the information connections provided by flows through, the compartments which comprise the ecosystem (Christensen 1995). Exergy and ascendancy are viewed as 'goal functions' of the system, aspects that are expected to be optimised or maximised by system development (Raffaelli and Frid, this volume).

If such measures are to be used to assess ecosystem health, then they should respond in a dose-dependent fashion to known stress (Jørgensen 1999, Zhang *et al.* 2003, Zhang *et al.* 2004), or differ between systems of known developmental stage (Christensen 1995, Baird *et al.* 1991, Wulff and Ulanowicz 1989, Christensen 1994). In his 1995 article, Christensen ranked the maturity of forty-one ecosystems and compared his ranking with that derived from maturity measures based on ascendancy and exergy. Maturity rank, assessed in terms of Odum's successional attributes, was strongly correlated with ascendancy. Exergy responded primarily to biomass rather than maturity, which led Christensen to suggest that alternative methods of calculating exergy might be more appropriate in future studies. Subsequent inclusion of the genetic complexity of organisms at different trophic levels within the exergy calculation has sought to correct this shortcoming (Jørgensen 1997, 1999, Jørgensen *et al.* 2007).

With such disparate assessment of ecosystem maturity arising from the model-derived system-level metrics proposed by researchers, it is unsurprising that conflicting results are frequently obtained when measures of this type are extended to indicators of ecosystem health (Dalsgaard *et al.* 1995, Lu and Li 2003, Xu *et al.* 2004). Despite these discrepancies, the concepts of ascendancy and exergy have received considerable attention within the ecosystem modelling and ecosystem health communities, and have been applied successfully for characterising estuarine ecosystems such as the Ythan catchment in Aberdeenshire (Raffaelli *et al.* 2005).

A full account of the historical changes within the catchment are provided in the text box (Box 4.1). Here, we consider energy-based approaches to assessing the health of the Ythan Estuary. Analyses of the Ythan catchment revealed that ascendancy increased by 50 per cent of the pre-eutrophic value following the onset of eutrophication, with system throughput changing by a similar factor (Raffaelli *et al.* 2005). System throughput is a measure of the total of all the flows and hence an expression of ecosystem 'size' (Kay *et al.* 1989). Both ascendancy and system throughput indices are consistent with the increase in biomass of the macroalgae *Enteromorpha* and the invertebrates *Hydrobia ulva*, *Macoma balthica* and *Nereis diversicolor*, which more than compensate for the tenfold decline in *Corophium volutator*, one of the main prey species of shorebirds. Because ascendancy reports the average mutual information of the system (its complexity) scaled by system throughput (Ulanowicz 1986), and the information measure of the Ythan is very similar in the 1960s and 1990s (*c.* 1.17), ascendancy in this ecosystem is driven mainly by system throughput. In other words, the system's basic food web structure, composition and topology (its complexity) are similar for the two periods, with no taxa going extinct, but the biomass and production of many elements are much higher following eutrophication. Relative ascendancy (expressed as a percentage of development capacity, a natural limit for ascendancy) was very similar in the two periods (*c.* 26 per cent) indicating that the Ythan *as a system* was able to accommodate the large-scale changes in nutrient loading, primary production and invertebrate biomass. In this sense, the Ythan eutrophication process is consistent with Ulanowicz's (1986) view that eutrophication can be described as any increase in system ascendancy due to nutrient enrichment that causes a rise in total system throughput, which more than compensates for any concomitant fall in the mutual information content.

Box 4.1 The Ythan Estuary

The Ythan Estuary in NE Scotland flows into the North Sea. Several major tributaries join the Ythan river towards its lower reaches, some almost as large as the main river itself, which is never more than a few tens of metres wide, even at the estuary. The estuarine ecosystem carries populations of nationally important flagship bird species including eider duck (*Somateria mollissima*), redshank (*Tringa totanus*), and shelduck (*Tadorna tadorna*), and has been studied intensively for more than 40 years by staff and students at the University of Aberdeen's Culterty Field Station. Here, we describe trends in indicators of environmental, ecological, financial, human and social capital stocks for the catchment as a whole.

Land use in the Ythan catchment is predominantly agricultural, and significant changes in agricultural practice have occurred within the area during the past 40 years, mirroring those elsewhere in Scotland (Raffaelli *et al.* 1989, 1999, 2005). The principal changes have been: the preferential growing of subsidised cereals, such as wheat and barley, at the expense of the more traditional oats; the introduction of novel crops, such as oilseed rape; an increase in the total land area under fertiliser-hungry cereals and rape, at the expense of grassland, especially rough grazing; a shift towards winter and autumn-sown cereals, such that land is tilled at a time of high precipitation and run-off; and an increase in pig production.

These land-use changes are unambiguously reflected in the water quality of the River Ythan. Since 1958, there has been a two- to three-fold increase in the concentration of total oxidised nitrogen (almost entirely nitrate) in river water (Raffaelli *et al.* 1989, Raffaelli *et al.* 1999) and a similar pattern is seen within the estuary. The Ythan catchment was consequently declared a Nitrate Vulnerable Zone (NVZ) under the European Union's Nitrates Directive and the UK is now required to take steps to reduce nitrate loadings within the river system to ameliorate the impact on the ecology of the estuary.

Whilst the eutrophication process in the Ythan is consistent with expectations from systems and network theory, this result serves to illustrate an ambiguity with the use of whole-system metrics for assessing ecosystem health. Over the last 40 years, the Ythan has displayed major changes in the populations of many species which were dramatic enough to see the catchment designated as a Nitrate Vulnerable Zone (NVZ) under the EC Nitrates Directive (Raffaelli *et al.* 1989, Raffaelli *et al.* 1999). The system-modelling approach has confirmed that the overall flows and biomasses have increased markedly, but system measures remain (relative to one another) broadly unchanged and, in absolute terms, ascendancy increased rather than declined. In other words, information that is of importance to stakeholders and policy makers on shifts in key and charismatic species is not necessarily captured by these system-level metrics. Indeed, it is possible that even a catastrophic collapse of the food web through the loss of shorebirds which would reduce the system's information content markedly, and hence potentially also its ascendancy, would be more than compensated for by enhanced algal growth and increased system throughput. From a systems-level perspective the Ythan could thus be considered resilient to a very significant external perturbation – nutrient enrichment. However, eutrophication effects are markedly non-monotonic and non-linear and as Raffaelli *et al.* (1989) and Raffaelli *et al.* (1999) have pointed out, continued increases in nutrient load and blooms of macroalgae would be expected to lead to collapse of the system.

Whilst mass-balance models may have potential for ecosystem health assessment, they are data intensive. Parameterisation of the Ythan model was possible because of the large body of work carried out over a 40-year period on the food web. Constructing similar ecosystem models of other systems would almost certainly require extensive data estimation or extrapolation from other studies such that confidence in the outputs might be correspondingly reduced.

Interdisciplinary indicators of ecosystem health

The HEHI approach

Many of the definitions and assessments of ecosystem health concentrate solely on biophysical aspects of the ecosystem (see above). It is, however, becoming increasingly recognised that the overall health of an ecosystem relies on a number of interrelated social, economic and ecological components. This highlights the importance of using a holistic approach that accommodates each of these components and addresses their interdependence, when assessing ecosystem health. Consequently, interdisciplinary indicators of ecosystem health are now being proposed which incorporate socio-economic, ecological, and community development components (e.g. Hannon 1992, Costanza 1994, Cobb *et al.* 1995).

One of the best-known examples of an interdisciplinary measure of ecosystem health is the Holistic Ecosystem Health Indicator (HEHI) (Aguilar 1999). HEHI incorporates ecological, social and interactive (interactions between human and ecological components) indicators to provide a more comprehensive assessment of the health of an ecosystem (ibid.). Each of the three components (ecological, social and interactive) is subdivided into categories depending on the ecological and social characteristics of the target area, and the management goals of the stakeholders involved (ibid.). The ecological component focuses on biophysical aspects of the ecosystem, particularly organisation, vigour and resilience (*sensu* Costanza 1992). The social component covers a range of socio-economic factors that are fundamental to the exploitation of ecosystem resources (Winograd 1995), and the indicators chosen within this category reflect the social and economic priorities of the communities which live in, or depend on, the ecosystem (Aguilar 1999). The interactive category quantifies the primary connections and relationships between people and the ecosystem, as well as the effectiveness of regulatory agencies in implementing legislation, community perceptions, awareness and involvement in management decisions (ibid.).

Specific indicators are selected to evaluate the condition of each of the three components, and ideally a benchmark is set for each of the indicators based on the scientific literature or management objectives and policy. A standardised scoring system is used to evaluate the individual indicators, with higher scores representing healthier ecosystems, and each indicator category is given a relative weighting depending on its importance to the overall health of the ecosystem and stakeholder goals (Muñoz-Erickson and Aguilar-González 2003). This approach was originally applied to tropical ecosystems in Costa Rica using nine ecological, six social, and six interactive indicator categories (Figure 4.1). In the Costa Rican study the ecological component was weighted at 40 per cent and both the other components at 30 per cent. Restricted availability of the

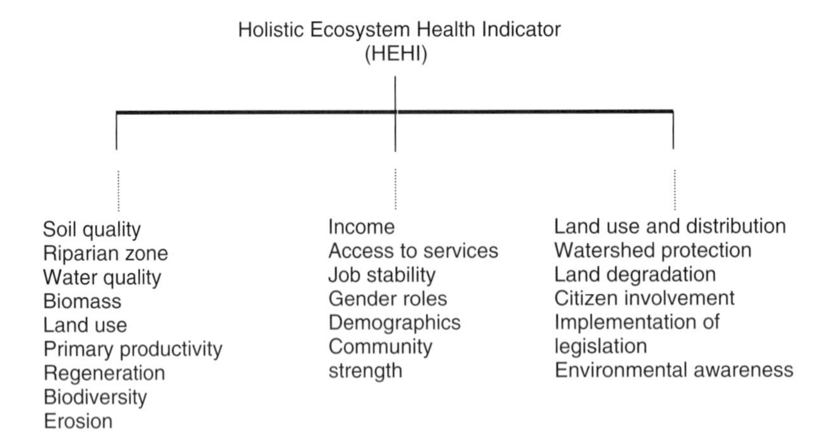

Figure 4.1 Structure of the Holistic Ecosystem Health Indicator (HEHI) used by Aguilar (1999) for tropical managed systems in Costa Rica.

necessary information only allowed a 'weak' health assessment to be produced as several indicators were lacking within each component (Aguilar 1999). HEHI was used in the Costa Rican situation to assess the health of the ecosystem at a single point in time, but it would be more informative for policy makers and managers to incorporate temporal trends into holistic approaches like HEHI in order to identify the rate and direction of any changes, as done for a novel method described below (MEHTA).

The main advantage of the HEHI approach is that it uses a simple and cost-effective methodology that allows managers and policy makers to focus their resources on the weakest aspects of ecosystem health. It also permits comparisons between different sites, and can thus reflect general trends at global, regional and local scales. Identifying appropriate indicators within each component, meaningful benchmarks for those indicators and a realistic timescale for assessment is, however, potentially difficult (Muñoz-Erickson *et al.* 2004). Value judgements have to be made (and defended), concerning the relative weightings of the three main components as well as in the directionality of the association between each of the indicators and ecosystem health. Ideally, these judgements are best made by local communities within existing regulatory and statutory frameworks. A drawback with composite indices such as HEHI is that they cannot unambiguously identify the underlying causes of changes in the health status of an ecosystem (Aguilar 1999).

MEHTA (Monitoring of Ecosystem Health by Trends Analysis): an alternative interdisciplinary indicator

In this section we take the holistic indicator approach a stage further. Specifically, we illustrate how the HEHI-type approach could be developed by incorporating a strong temporal component, ideally incorporating historical data, to assess the rate of change in individual indicators and their directionality with respect to targets or thresholds established by a combination of statutory requirements and stakeholder consultation.

Our guiding principle in developing this new indicator has been that the maintenance of healthy ecosystems is a prerequisite for sustainable development and thus a good indicator of ecosystem health should represent a measure of social welfare or utility. Certain products of value to society arise from stocks of environmental and ecological capital with minimal human intervention (e.g. rare or charismatic species and their associated existence values, clean air or water delivered by the natural purification services afforded by forests and woodlands), and system resilience to invasive species and disease. Other products of socio-economic relevance arise only when ecosystem services underpinned by stocks of ecological and environmental capital are combined with stocks of man-made capital (financial, human or social capital) through active management of the ecosystem concerned. For example, agricultural crops are produced by combining the pollination and soil fertility services, which are supported by ecological and environmental capital, with human, financial and social capital in the form of agricultural labour, investment in seed and equipment provision and the sales and marketing infrastructure of agri-business. Thus, in order to ensure that managed ecosystems continue to deliver a desired bundle of products of relevance and value to society, adequate stocks of environmental, ecological, financial, human and social capital must be maintained within those managed ecosystems.

The health status of a managed ecosystem could therefore be assessed by monitoring the status of the environmental, ecological, financial, human and social capital stocks associated with that ecosystem, relative to critical thresholds for those stocks which are necessary to maintain delivery of a desired bundle of products. The approach developed here (Monitoring Ecosystem Health by Trends Analysis; MEHTA) uses an appropriate set of indicators to report on the status of these underlying capital stocks, assesses the safety margins which remain before the critical thresholds for each stock are infringed, and derives a measure of the rate at which capital stocks are being depleted or enhanced by utilising time series data for the individual indicators concerned.

Weightings are produced by determining the relative socio-economic value attached to products derived from the different capital stocks. These relative valuations would be ideally elicited from the social community which is part of the ecosystem concerned. The MEHTA approach thus has the potential to incorporate stakeholder knowledge and preferences to produce a relative valuation of products derived from the ecosystem. Expert knowledge can then be applied to determine those elements of natural and man-made capital which support product delivery and to establish critical thresholds for the capital stocks concerned. This approach to the assessment of ecosystem health incorporates the values and aspirations of society together with expert knowledge of ecosystem structure and function, and thereby yields a health assessment which can be regarded as a prerequisite for sustainable development.

Indicator data are generally available as time series measurements, detailing, for example, nitrate levels in river water, number of visitors to an area, abundance of a particular bird species, or average income per household. The advantage of utilising time series data is that statistically significant trends can be identified and quantified which allows the rate of approach to prescribed critical thresholds to be estimated (Figure 4.2). A health weighting can be assigned to the trend and safety margin results for each indicator based on the functional importance of the capital stock concerned and the relative valuations placed by society on those products. The health implications for each element of capital stock within the overall health assessment are scaled by the weighting factor for that indicator to produce a health score. Scores can then

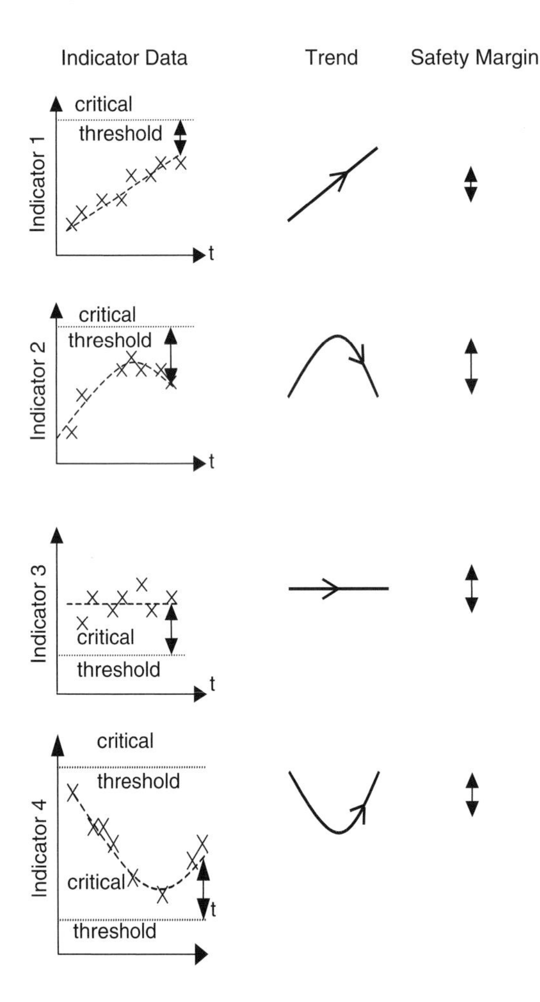

Figure 4.2 Assessing ecosystem health by determining trends and safety margins for individual indicators of the status of environmental, ecological, financial, human and social capital stocks within a managed ecosystem. An illustration of the MEHTA (Monitoring of Ecosystem Health by Trends Analysis) approach.

either be summed across the individual indicators to produce a composite health score for the ecosystem as a whole or, perhaps more informatively, be presented as a series of individual trends.

The MEHTA approach has some parallels with HEHI (Aguilar 1999), but there are two important differences. First, MEHTA indicators reflect the status of an essential set of environmental, ecological, financial, human and social capital stocks which underpin the provision of a desired bundle of products within the managed ecosystem. In this sense, they are derived from first principles. Second, the analysis utilises historical time series data to determine the rate of approach to critical thresholds associated with indicators of separate elements of capital stock, which enables the (statistical) uncertainty surrounding these trends and safety margins to be quantified. The latter feature should also permit quantification of any uncertainty surrounding future predictions of ecosystem status.

Application of the MEHTA approach: Ythan catchment case study

Here, we illustrate the potential and limitations of the MEHTA approach by exploring its application to the Ythan catchment in Aberdeenshire, before comparing the results with the conclusions from the mass-balance approach previously described. The Ythan is a lowland Scottish catchment, *c.* 640 km^2 in area, which rises to a few hundred metres in altitude to the north of the city of Aberdeen in the north-east of Scotland. Ninety percent of the catchment is under agriculture, and the pressures and drivers of ecosystem change in the catchment are extremely well-documented and understood. It is not our intention here to provide a definitive statement about the health of the Ythan catchment systems, but rather to show how the MEHTA approach might be usefully applied. A full assessment of the health of the system would require more extensive data sets and the use of participatory research techniques with stakeholders to define acceptable thresholds and limits.

In applying the MEHTA approach to assess the health of the Ythan catchment, we regarded the following capital stocks as central to the continued delivery of a wide range of products of relevance to the community:

1. Ecological capital – in the form of biodiversity, which supports ecosystem services such as soil fertility, natural pest control and pollination, which underpin delivery of agricultural crops, and also maintain landscape features which attract tourists and visitors to the catchment. The abundance of wader birds recorded on the Ythan estuary (Raffaelli *et al.* 1999), and an index of the abundance of breeding birds (Raven *et al.* 2004) were used as indicators of biodiversity as a stock of ecological capital within the catchment. These indicators report the status of different bird species and so are regarded here as surrogates for different aspects of biodiversity within the

catchment. The breeding bird survey covered the whole of the Grampian region as data were not available at finer spatial resolution.

2. Environmental capital – in the form of the water purification capacity provided by the catchment. An index of water quality in the Ythan (Scottish Environmental Protection Agency (SEPA), unpublished) was used as a measure of aquatic ecosystem health, and as a surrogate for available water purification capability as a stock of environmental capital.
3. Human and social capital – in the form of an appropriately skilled labour force accessible to land management businesses, viable rural communities and functional rural infrastructure for the production and dissemination of products generated by land management businesses, including businesses which service tourism. The population living within the catchment (Aberdeenshire Council, personal communication), and the number of those that were employed (Office of National Statistics, personal communication) were used as indicators of the stocks of human and social capital within the catchment.

Trends within the data, together with their associated confidence intervals (95 per cent), were analysed using simple and multiple linear regression. Simple and multiple linear regressions were performed on the time series data sets for each indicator using SPSS (v11). The data were checked for outliers and influential cases using standardised residuals and Cook's distance, respectively. Explanatory variables were introduced sequentially (manually stepwise) into the multiple linear regressions and only those variables that significantly improved the fit of the model were retained. The trend results obtained from the regression analyses for individual indicators are shown in Table 4.4.

The mean abundance of wading birds recorded on the Ythan estuary followed a second order polynomial curve, peaking in the early 1980s. The decline recorded after this point may be a response to the negative impact which changing patterns of agriculture within the catchment exerted on water quality (Raffaelli *et al.* 1999, 2005). Data detailing the abundance of breeding birds also followed a second-order polynomial curve with a minimum occurring in 1998. These data were only available from 1994–2003, and it was therefore not possible to compare breeding bird abundance pre- and post-eutrophication. The current increase in abundance may be a result of recent conservation initiatives, for example NVZ and ESA (Environmentally Sensitive Areas) measures.

Water quality in the Ythan estuary decreased significantly between 1980 and 1990, reflecting increased fertiliser use and slurry application within the catchment. This matches the explanation given in the preceding section regarding

Table 4.4. *Summary of trends in indicators of natural and man-made capital stocks in the Ythan catchment.*

Capital Stock	Indicator	Time span	Model Fit	R^2 (adj. R^2)	Trend
Ecological	Breeding birds	1994–2003 (excl. 2001)	$F_{2,6}$=5.203 P=0.049	0.634 (0.482)	
	Wading birds	1969, 1970, 1973–1978, 1980–1982, 1988–1995	$F_{2,16}$=5.62 P=0.014	0.413 (0.212)	
Environmental	Water quality	1980–1990 (excl.1985)	$F_{1,8}$=17.850 P=0.003	0.691 (0.563)	
Social and Human	Population	1901–2001 (every decade excl.1941)	$F_{2,7}$=10.029 P=0.009	0.741 (0.512)	
	Employment	1984, 1987, 1989, 1991, 1993, 1995–1998	$F_{3,5}$=8.665 P=0.02	0.839 (0.499)	

the decrease in wading bird numbers recorded on the estuary over this time period.

The human population within the catchment followed a second-order polynomial curve with a minimum around the early 1950s. The population has increased and major changes in land use have occurred within the catchment since the late 1960s. Many factors may have contributed to this increase, including the boom in the oil industry across Aberdeenshire as a whole. Employment figures within the catchment showed no significant change between 1984 and 1991, but, post-1991, the number of people in employment increased significantly.

MEHTA revealed that indicators of ecological and environmental capital within the Ythan catchment declined significantly with the onset of eutrophication (up to 1990). The decline in wading birds and water quality reflected a reduction in the health of the aquatic ecosystems within the catchment, probably as a consequence of increased nitrate run-off into the river system caused by fertiliser and slurry application. The increase in population, however, suggests that the area became more desirable towards the end of the 1980s, but this may be a result of the employment generated within the oil industry during this period. The overall conclusions for the health of managed ecosystems within the Ythan catchment are therefore that stocks of social and human capital appear to have increased whilst ecological and environmental capital stocks have declined.

Conclusions and recommendations

Is the ecosystem health concept valuable?

Understanding and assessing ecosystem health are important because ecosystem health underpins sustainable development. However, because of the widespread and increasing impacts of humans throughout the world, ecosystem health is of limited use as a concept when it is applied to the non-human biological components of a system in isolation. For this reason, ecosystem health as presented here encompasses the environmental, economic and social dimensions as well, to provide a more holistic assessment of sustainability, one which resonates well with the CBD's Malawi Principles (Frid and Raffaelli, this volume). It also provides a means through which society's views on ecosystems and the environment can be formally incorporated into this assessment via participatory approaches. This is not to say that the views of society should entirely supplant expert judgements, but they should be included as far as possible in any overall evaluation of sustainability and conservation issues.

The measurement of ecosystem health also has the potential to highlight heterogeneities in the way that stakeholders in different areas value stocks of the different forms of natural and man-made capital within managed ecosystems. Such differences may produce geographical differences in the outcome of an ecosystem health assessment.

Operational approaches to assessing ecosystem health

Many of the earlier proposals for ecosystem health indices, as discussed in the previous chapters, are primarily useful as conceptual rather than operational models (e.g. Costanza 1992). Other models are dependent on detailed measurements from food webs that can only come from intensive, long-term empirical studies (e.g. Hannon 1985, Ulanowicz 1992, Jørgensen 1995). More recent work highlighting the importance of humans in an assessment of ecosystem health (Xu and Mage 2001) has extended the assessment criteria in conceptual terms, but the only operational tool to be used prior to the development of MEHTA is HEHI (Aguilar 1999).

The MEHTA approach differs from the HEHI approach in that HEHI is an assessment based on the values of specific indicators at one point in time, whereas MEHTA is explicitly based on an assessment of trends in indicators of stocks of natural and man-made capital over time. It therefore provides a means by which changes in ecosystem health can be monitored, and allows the rate and direction of this change to be evaluated with respect to specific critical thresholds. Both approaches allow participatory involvement along with expert knowledge to determine the weightings attached to various indicators, which is an important criterion in any health assessment.

The application of MEHTA to the Ythan catchment provides a much more complete understanding of the interactions between social, environmental and ecological capital than the purely ecological insights provided by the mass-balance approach described earlier. The analysis was able to reveal apparent trade-offs between environmental and ecological versus human and social capital over time. This type of integrated approach for assessing ecosystem health has direct application to the development of policy regarding sustainable development in specific areas, and moreover provides an opportunity for stakeholders and the public to become actively involved in determining preferences in relation to policy development.

Gaps in knowledge

Ecosystem health is a relatively young subject, and paradigm shifts in terms of understanding may occur. The interdisciplinary nature of the subject has led to very different approaches being taken, and there is no generally accepted methodology for evaluating ecosystem health, or even a universally accepted definition of its scope. Nevertheless, one key area where there are major gaps in understanding is the relationship between social and human capital, and the other three stocks of capital (financial, ecological and environmental) within ecosystems and in particular the role of social and human capital in promoting conservation, and/or use of these other stocks. HEHI and MEHTA provide examples of initial mechanisms for integrating these different stocks, but more work needs to be done, especially in relation to cause–effect relationships and feedbacks among the different capital stocks.

Mechanisms by which the relative values which society ascribes to stocks of financial, environmental and ecological capital, and the way in which these relative values can be incorporated into the formal ecosystem health assessment, should also be a focus of research. Understanding the weightings of the different components in relation to social capital and governance is a critical issue for policy makers in terms of conserving ecosystem health and enhancing sustainability. Several of these aspects are discussed further in Haines-Young and Potschin (this volume).

Many of the indicator monitoring systems currently used focus on specific ecological, environmental or social elements which are sometimes difficult to relate to the overall health of the broader system in which all these different elements are embedded and interact. In addition, the relationship between many of these measures to capital stocks and ecosystem services is not always articulated or clear. If environmental quality and social well-being are to be optimised, then the ecosystem-health approach, which is holistic and has measures unambiguously related to capital stocks and services, offers considerable potential.

Implementing this approach operationally will require a shift in thinking at political and scientific levels. However, there are indications that many central governments are increasingly recognising the importance of interrelationships between socio-economic and environmental components in ecosystems and starting to embrace a more holistic approach to policy making. Human behaviour is the main driver of environmental change, so humans need to be seen as part of, not apart from, the ecosystem. Maintaining human well-being will require monitoring of natural and social capital stocks and the interactions between them. In this respect, holistic indicator systems, such as HEHI and MEHTA described here, offer considerable potential for evaluating the health of coupled social–ecological systems. However, it is important that these measures are underpinned by a solid foundation of data across the social and natural sciences, and that these data can be integrated using appropriate spatial and temporal units. Policies focused on individual environmental or social elements will be unlikely to protect underlying stocks, the services which flow from them and human well-being.

Perhaps most challenging politically will be the derivation of acceptable levels to society of stocks of natural capital. If policies are to be successfully implemented with respect to the regulation of natural capital, so that the desired ecosystem services can be maintained, then there must be both scientific and social inputs into those discussions which set those thresholds.

Acknowledgements

Much of this material is based on work supported by the UK's Department of Environment, Food and Rural Affairs (Defra), for which we are extremely grateful.

References

Abarca-Arenas, L.G. and Ulanowicz, R.E. (2002) The effects of taxonomic aggregation on network analysis. *Ecological Modelling*, **149**, 285–96.

Aguilar, B.J. (1999) Applications of ecosystem health for the sustainability of managed systems in Costa Rica. *Ecosystem Health*, **5**, 1–13.

Ali, A.M.S. (1995) Population pressure, environmental constraints and agricultural change in Bangladesh: examples from three agroecosystems*1. *Agriculture, Ecosystems & Environment*, **55**, 95–109.

Baird, D., McGlade, J.M. and Ulanowicz, R.E. (1991) The comparative ecology of six marine ecosystems. *Philosophical Transactions of the Royal Society (London)*, **333**, 15–29.

Bartels, G. and Kampmann, T. (1994) Auswirkungen eines langjährigen Einsatzes von Pflanzenschutzmittel bei unterschiedlichen Intensitätsstufen und Entwicklung von Bewertungskriterien. *Mitteilungen aus der Biologischen Bundeanstalt für Land- und Forstwirtschaft*, **295**, 1405.

Belaoussoff, S. and Kevan, P.G. (1998) Toward an ecological approach for the assessment of ecosystem health. *Ecosyst Health*, **4**, 4–8.

Bockstaller, C. and Girardin, P. (2003) How to validate environmental indicators. *Agricultural Systems*, **76**, 639–53.

Bockstaller, C., Girardin, P. and van der Werf, H.M.G. (1997) Use of agro-ecological indicators for the evaluation of farming systems. *European Journal of Agronomy*, **7**, 261–70.

Boulton, A.J. (1999) An overview of river health assessment: philosophies, practice, problems and prognosis. *Freshwater Biology*, **41**, 469–79.

Buchs, W. (2003) Biodiversity and agri-environmental indicators – general scopes and skills with special reference to the habitat level. *Agriculture, Ecosystems & Environment*, **98**, 35–78.

Christensen, V. (1994) On the behavior of some proposed goal functions for ecosystem development. *Ecological Modelling*, **75/76**, 37–49.

Christensen, V. (1995) Ecosystem maturity – towards quantification. *Ecological Modelling*, **77**, 3–32.

Christensen, V. and Walters, C.J. (2004) Ecopath with Ecosim: methods, capabilities and limitations. *Ecological Modelling*, **172**, 109–39.

Cobb, C., Halstead, T. and Rowe, J. (1995) *The genuine progress indicator: summary of data and methodology.* Redefining Progress, San Francisco, California.

Conway, G.R. (1985) Agroecosystem analysis (Thailand). *Agricultural Administration*, **20**, 31–55.

Conway, G.R. (1987) The properties of agro-ecosystems. *Agricultural Systems*, **24**, 95–117.

Costanza, R. (1992) Toward an operational definition of ecosystem health. In *Ecosystem Health:New Goals for Environmental Management* (eds. R. Costanza, B.G. Norton and B.D. Haskell). Island Press, Washington, DC.

Costanza, R. (1994) Ecological and economics system health and social decision making. *1st International Symposium on Ecosystem Health and Medicine: Integrating Science, Policy and Management*, pp.24. International Society for Ecosystem Health and Medicine, University of Guelph, Ottawa, Ontario.

Dalsgaard, J.P.T., Lightfoot, C. and Christensen, V. (1995) Towards quantification of ecological sustainability in farming systems analysis. *Ecological Engineering*, **4**, 181–9.

Duelli, P. and Obrist, M.K. (2003) Biodiversity indicators: the choice of values and measures. *Agriculture, Ecosystems & Environment*, **98**, 87–98.

Epstein, P.R. and Rapport, D.J. (1996) Changing coastal marine environments and human health. *Ecosystem Health*, **2**, 166–76.

Ferris, R. and Humphrey, J.W. (1999) A review of potential biodiversity indicators for application in British forests. *Forestry*, **72**, 313–28.

Fisheries Centre University of British Columbia. (2004) Ecopath website.

Gallopin, G.C. (2002) Planning for resilience: Scenarios, surprises and branch points. In *Panarchy: Understanding transformations in human and natural systems* (eds. L.H. Gunderson and C.S. Holling). Island Press, Washington, DC.

Gomez, A.A., Kelly, D.E., Syers, J.K. and Coughlan, K.J. (1996) Measuring sustainability of agricultural systems at the farm level. *Methods for Assessing Soil Quality, SSSA Special Publication*, **49**, 401–9.

Gunderson, L.H. and Holling, C.S. (eds.) (2002) *Panarchy: Understanding Transformations in Human and Natural Systems.* Island Press, Washington, DC.

Gunderson, L.H. and Pritchard, L. (2002) *Resilience and the Behavior of Large Scale Ecosystems.* Island Press, Washington, DC.

Hannon, B. (1985) Ecosystem flow analysis. *Canadian Journal of Fisheries and Aquatic Sciences*, **213**, 97–118.

Hannon, B. (1992) Measures of economic and ecological health. In *Ecosystem health: New Goals for Environmental Management* (eds. R. Costanza, B.G. Norton and B.J. Haskell), pp.207–22. Island Press, Washington, DC.

Haskell, B.D., Norton, B.G. and Costanza, R. (1992) Introduction: What is ecosystem health and why should we worry about it? In *Ecosystem Health: New Goals for Environmental Management* (eds. R. Costanza, B.G. Norton and B.D. Haskell), pp.3–20. Island Press Washington, DC.

Hoffmann, J. and Greef, J.M. (2003) Mosaic indicators - theoretical approach for the development of indicators for species diversity in agricultural landscapes. *Agriculture, Ecosystems & Environment*, **98**, 387–94.

Hoffmann, J., Greef, J.M., Kiesel, J., Lutze, G. and Wenkel, K.-O. (2003) Practical example of the mosaic indicators approach. *Agriculture, Ecosystems & Environment*, **98**, 395–405.

Holland, J.M., Frampton, G.K., Cilgi, T. and Wratten, S.D. (1994) Arable acronyms analysed - a review of integrated arable farming systems research in western Europe. *Annals of Applied Biology*, **125**, 399–438.

Holling, C.S. (1973) Resilience and stability of ecological systems. *Annual Review of Ecology and Systematics*, **4**, 1–23.

Holling, C.S. (1986) The resilience of terrestrial ecosystems: local surprise and global change. In *Sustainable Development of the Biosphere* (eds. W.C. Clark and R.E. Munn), pp.292–317. Cambridge University Press.

Holling, C.S. (1992) Cross-scale morphology, geometry and dynamics of ecosystems. *Ecological Monographs*, **62**, 447–502.

Holling, C.S., Gunderson, L.H. and Peterson, G.D. (2002) Sustainability and Panarchies. In *Panarchy: Understanding Transformations in Human and Natural Systems* (eds. L.H. Gunderson and C.S. Holling), pp.63–102. Island Press, Washington, DC.

Huq, A. and Colwell, R.R. (1996) Vibrios in the marine and estuarine environment: tracking Vibrio cholerae. *Ecosystem Health*, **2**, 198–214.

Johnson, G.D., Brooks, R.P., Myers, W.L., O'Connell, T.J. and Patil, G.P. (2002) Predictability of bird community-based ecological integrity, using landscape measurements. In *Managing for Healthy Ecosystems* (eds. D. Rapport, W. Lasley, D. Rolston, O. Neilson, C. Qualset and A. Damania). CRC Press Boca Raton, FL.

Jørgensen, S.E. (1986) Structural dynamic model. *Ecological Modelling*, **31**, 1–9.

Jørgensen, S.E. (1988a) *Fundamentals of Ecological Modelling*. Elsevier, Amsterdam, the Netherlands.

Jørgensen, S.E. (1988b) Use of models as experimental tool to show that structural changes are accompanied by increase in exergy. *Ecological Modelling*, **41**, 117–126.

Jørgensen, S.E. (1992) Development of models able to account for changes in species composition. *Ecological Modelling*, **62**, 195–208.

Jørgensen, S.E. (1995) Exergy and ecological buffer capacities as measures of ecosystem health. *Ecosystem Health*, **1**, 150–60.

Jørgensen, S.E. (1997) *Integration of Ecosystem Theories: A Pattern*. Kluwer, Dordrecht, the Netherlands.

Jørgensen, S.E. (1999) State-of-the-art of ecological modelling with emphasis on development of structural dynamic models. *Ecological Modelling*, **120**, 75–96.

Jørgensen, S.E., Nielsen, S.N. and Mejer, H. (1995) Emergy, environ, exergy and ecological modelling. *Ecological Modelling*, **77**, 99–109.

Jørgensen, S.E., Fath, B.D., Bastianoni, S., Marques, J.C., Muller, F., Nielson, S.N, Patten, B.C., Tiezzi, E. and Ulanowicz, R.E. (2007) *A New Ecology. Systems Perspective*. Elsevier, Amsterdam, the Netherlands, 275.

Karr, J.R. (1992) Ecological integrity: protecting earth life support systems. In *Ecosystem Health: New Goals for Environmental Management* (eds. R. Costanza, Norton, B.G., Haskell, B.D.), pp.228–38. Island Press, Washington, DC.

Karr, J.R. (1999) Defining and measuring river health. *Freshwater Biology*, **41**, 221–34.

Karr, J.R., Fausch, K.D., Angermeier, P.L., Yant, P.R. and Schlosser, I.G. (1986) *Assessing Biological Integrity in Running Waters: A Method and its Rationale*. Illinois Natural History Survey; Special Publication 5, Champaign, Illinois, USA.

Kay, J.J., Graham, L. and Ulanowicz, R.E. (1989) A Detailed Guide to Network Analysis. In *Network Analysis in Marine Ecology: Methods and Applications* (eds. F. Wulff, J. Field and K. Mann). Springer-Verlag, Berlin.

Lu, F. and Li, Z. (2003) A model of ecosystem health and its application. *Ecological Modelling*, **170**, 55–9.

Muñoz-Erickson, T.A. and Aguilar-González, B.J. (2003) The use of ecosystem health indicators in evaluating ecological and social outcomes of collaborative approaches to management: the case study of the Diablo Trust. In *Evaluating methods and environmental outcomes of community based collaborative processes*. Snowbird Center, Salt Lake City, UT.

Muñoz-Erickson, T.A., Loeser, M.R. and Aguilar-González, B.J. (2004) Identifying indicators of ecosystem health for a semiarid ecosystem: a conceptual approach. In *The Colorado Plateau: cultural, biological and physical research* (eds. C.I. van Riper and K.L. Cole). University of Arizona Press, Tucson, AZ.

Nielsen, S.N. and Ulanowicz, R.E. (2000) On the consistency between thermodynamical and network approaches to ecosystems. *Ecological Modelling*, **132**, 23–31.

Odum, E.P. (1953) *Fundamentals of Ecology*, 1st edn. W.B. Saunders, Philadelphia.

Odum, E.P. (1969) The strategy of ecosystem development. *Science*, **164**, 262–70.

Odum, E.P. (1985) Trends expected in stressed ecosystems. *Bioscience*, **35**, 419–22.

Parsons, M. and Norris, R.H. (1996) The effect of habitat-specific sampling on biological assessment of water quality using a predictive model. *Freshwater Biology*, **36**, 419–34.

Patil, G.P. (2000) Multiscale advanced raster map analysis for sustainable environment and development: A research and outreach prospectus of advanced mathematical, statistical and computational approaches using remote sensing data. Development and implementation of a prototype marmap remote sensing application, technology and education for multiscale advanced raster map analysis program. The Pennsylvania State University.

Patil, G.P. (2001a) Biocomplexity of ecosystem health and its measurement at the landscape scale. A research and outreach prospectus of advanced mathematical, statistical and computational approaches using remote sensing data and GIS. Development and implementation of a prototype marmap. Remote sensing application, technology and education for multiscale advanced raster map analysis program for biocomplexity of ecosystem health and its measurement at the landscape scale. The Pennsylvania State University.

Patil, G.P. (2001b) Cost effective ecological synthesis and environmental analysis research and outreach: A prospectus. The Pennsylvania State University.

Patil, G.P. (2001c) Linkage of multiscale multisource multi-tier data for the purposes of regional assessments and monitoring: A research and outreach prospectus of advanced mathematical, statistical, and computational approaches. The Pennsylvania State University.

Patil, G.P., Brooks, R.P., Myers, W.L., Rapport, D.J. and Taillie, C. (2002) Ecosystem health and its measurement at landscape scale: towards the next generation of quantitative assessments. *Ecosystem Health*, **7**(4), 307–16.

Patil, G.P. and Myers, W.L. (1999) Editorial: Environmental and ecological health assessment of landscapes and watersheds with remote sensing data. *Ecosystem Health*, **5**, 221–4.

Piorr, H.-P. (2003) Environmental policy, agri-environmental indicators and landscape indicators. *Agriculture, Ecosystems & Environment*, **98**, 17–33.

Raffaelli, D., Hull, S. and Milne, H. (1989) Long-term changes in nutrients, weed mats and shorebirds in an estuarine system. *Cahiers de Biologie Marine*, **30**, 259–70.

Raffaelli, D.G., Balls, P., Way, S., Patterson, I.J., Hohman, S.A. and Corp, N. (1999) Major changes in the ecology of the Ythan estuary, Aberdeenshire: how important are physical factors? *Aquatic Conservation: Marine and Freshwater Ecosystems*, **9**, 219–36.

Raffaelli, D., White, P., Renwick, A, Smart, J.C.R. and Perrings, C. (2005) The Health of Ecosystems: The Ythan Estuary Case Study. In *Handbook of Indicators for Assessment of Ecosystem Health* (eds. S.E. Jørgensen, R. Costanza and F-L. Xu,) CRC Press, Boca Raton, FL, USA.

Rapport, D., Regier, H.A. and Hutchinson, T.C. (1985) Ecosystem behaviour under stress. *American Naturalist*, **125**, 617–40.

Rapport, D.J. (1995) Ecosystem health: Exploring the territory. *Ecosystem Health*, **1**, 5–13.

Rapport, D.J., Bohm, G., Buckingham, D., Cairns, J., Jr., Costanza, R., Karr, J.R., de Kruijf, H.A.M., Levins, R., McMicheal, A.J., Neilson, N.O. and Whitford, W.G. (1999) Ecosystem Health: the concept, the ISEH, and the important tasks ahead. *Ecosystem Health*, **5**, 82–90.

Raven, M.J., Noble, D.G. and Baille, S.R. (2004) The Breeding Bird Survey 2003. BTO Research Report 363. *British Trust for Ornithology*, Thetford.

Reynoldson, T.B., Bailey, R.C., Day, K.E. and Norris, R.H. (1995) Biological guidelines for freshwater sediment based on BEnthic Assessment of Sediment (the BEAST) using a multivariate approach for predicting biological state. *Australian Journal of Ecology*, **20**, 198–219.

Rigby, D., Woodhouse, P., Young, T. and Burton, M. (2001) Constructing a farm level indicator of sustainable agricultural practice. *Ecological Economics*, **39**, 463–78.

Robinson, L.A. and Frid, C.L.J. (2003) Dynamic ecosystem models and the evaluation of ecosystem effects of fishing: can we make meaningful predictions? *Aquatic Conservation: marine and freshwater ecosystems*, **13**, 5–20.

Ruesink, J.L., Hodges, K.E. and Krebs, C.J. (2002) Mass-balance analyses of boreal forest population cycles: Merging demographic and ecosystem approaches. *Ecosystems*, **5**, 138–58.

Steinmann, H.-H. and Gerowitt, B. (2000) Ackerbau in der Kulturlandschaft-Funktionen und Leistungen. *Ergebnisse des Göttinger INTEX-Projecktes*, pp. 300. Mecke Verlag, Duderstadt, Germany.

Stork, N.E. (1995) Measuring and monitoring arthropod diversity in temperate and tropical forests. In *Measuring and monitoring biodiversity in temperate and tropical forests* (eds. T.J.B. Boyle and B. Boontawee), 257–70. CIFOR, Bogor, Indonesia.

Stork, N.E. and Eggleton, P. (1992) Invertebrates as determinants and indicators of soil quality. *American Journal of Alternative Agriculture*, **7**, 38–47.

Sueter, G.W. (1993) A critique of ecosystem health concepts and indices. *Environmental Toxicology and Chemistry*, **12**, 1533–9.

Swete-Kelly, D. (1996) Development and evaluation of sustainable systems for steeplands – lessons for the South Pacific. *Sustainable land management in the South Pacific*. IBSRAM, Thailand.

Ulanowicz, R.E. (1980) An hypothesis on the development of natural communities. *Journal of Theoretical Biology*, **85**, 223–45.

Ulanowicz, R.E. (1986) *Growth and Development: Ecosystem Phenomenology*. Springer-Verlag, New York, USA.

Ulanowicz, R.E. (1992) Ecosystem health and trophic flow networks. In *Ecosystem Health: New Goals for Environmental Management* (ed. B.D. Haskell). Island Press, Washington, DC.

Ulanowicz, R.E. and Kay, J.J. (1991) A package for the analysis of ecosystems flow networks. *Environmental Software*, **6**(3), 131–42.

van der Werf, H.M.G. and Petit, J. (2002) Evaluation of the environmental impact of agriculture at the farm level: a comparison and analysis of 12 indicator-based methods. *Agriculture, Ecosystems & Environment*, **93**, 131–45.

Waldhardt, R., Simmering, D. and Albrecht, H. (2003) Floristic diversity at the habitat scale in agricultural landscapes of Central Europe – summary, conclusions and perspectives. *Agriculture, Ecosystems & Environment*, **98**, 79–85.

Walker, J. (2002) Environmental indicators and sustainable agriculture. In *Regional Water and Soil Assessment for Managing Sustainable Agriculture in China and Australia* (eds. T. McVicar, L. Rui, J. Walker, R.W. Fitzpatrick and L. Changming). ACIAR, Canberra, Australia, pp. 323–32.

Waltner-Toews, D. (1994) Ecosystem health: a framework for implementing sustainability in agriculture. In *Proceedings of an International Workshop on Agroecosystem Health* (ed N.O. Nielsen), pp. 8–23. University of Guelph, Guelph, Canada.

Wichert, G. and Rapport, D.J. (1998) Fish community structure as a measure of degradation and rehabilitation of riparian systems in an agricultural drainage basin. *Environmental Management*, **22**, 425–43.

Winograd, M. (1995) *Indicadores ambientales para America Latina y el Caribe: hacia la sustenabilidad en el uso de tierras*. IICA, San Jose, Costa Rica.

Wright, J.F. (1995) Development and use of a system for predicting the macroinvertebrate fauna in flowing waters. *Australian Journal of Ecology*, **20**, 181–97.

Wulff, F. and Ulanowicz, R.E. (1989) A comparative anatomy of the Baltic Sea and Chesapeake Bay ecosystems. In *Network Analysis in Marine Ecology – Methods and Applications* (eds. F. Wulff, J.G. Field and K.H. Mann). Springer-Verlag, New York, USA.

Xu, F.-L., Jørgensen, S.E. and Tao, S. (1999) Ecological indicators for assessing freshwater ecosystem health. *Ecological Modelling*, **116**, 77–106.

Xu, F.L., Lam, K.C., Zhao, Z.Y., Zhan, W., Chen, Y.D. and Tao, S. (2004) Marine coastal ecosystem health assessment: a case study of the Tolo Harbour, Hong Kong, China. *Ecological Modelling*, **173**, 355–70.

Xu, W. and Mage, J.A. (2001) A review of concepts and criteria for assessing agroecosystem health including a preliminary case study of southern Ontario. *Agriculture, Ecosystems & Environment*, **83**, 215–33.

Zhang, J.J., Jørgensen, S.E. and Mahler, H. (2004) Examination of structurally dynamic eutrophication model. *Ecological Modelling*, **173**, 313–33.

Zhang, J.J., Jørgensen, S.E., Tan, C.O. and Beklioglu, M. (2003) A structurally dynamic modelling – Lake Mogan, Turkey as a case study. *Ecological Modelling*, **164**, 103–20.

CHAPTER FIVE

Interdisciplinarity in ecosystems research: developing social robustness in environmental science

KEVIN EDSON JONES
Management School, University of Liverpool
ODETTE A. L. PARAMOR
School of Environmental Sciences, University of Liverpool

Introduction

Within the academic community, there is a strong rhetorical value surrounding the idea of interdisciplinarity in research and teaching, but, as is often the case, that rhetoric about the benefits of collaboration outpaces developments in practice (Huber 1992). It is not that research which crosses disciplinary boundaries does not exist; interdisciplinary subgroups and research centres abound in academic institutions. Research councils, likewise, are progressively making interdisciplinarity more and more part of the core criteria by which funding is allocated to the research community. However, despite flurries of activity, the reality of engagement can be less satisfying and achieve less than initially envisioned (Pickett *et al.* 1999, Tress *et al.* 2005, Raffaelli and Frid this volume). Yet, the urgent need to better understand the complex environmental problems facing society and the imperative to find solutions to these problems remain and continue to compel the development of interdisciplinary innovation in environmental studies.

In this chapter we explore the development, promises and challenges of research which crosses the boundaries between the ecological, social and economic sciences, and what this means for the development of ecosystems research. We aim to reinforce and give depth to rationales for collaboration as well as arguing for a more ambitious and reflexive approach to interdisciplinarity and, hopefully, to provide impetus for a more meaningful and satisfactory experience for those taking the interdisciplinary road.

Underlying this endeavour is the assumption that addressing complex shared environmental problems initially requires researchers to draw together a wider range of knowledge than could be delivered by any single discipline. However, we argue that for collaboration to be effective, it is essential that intellectual reflection is allowed to enable researchers to develop novel ideas and synergies (Kinzig 2001). While interdisciplinarity is not necessarily an

Ecosystem Ecology: A New Synthesis, eds. David G. Raffaelli and Christopher L. J. Frid. Published by Cambridge University Press.

exclusive prerequisite to good ecosystems research, we propose that it can be beneficial in reflexively developing the discipline and in making outputs meaningful when addressing environmental problems. With this end in mind, this chapter first explores how developing an understanding of ecosystems research in a social context can produce more socially robust and reflexively aware research. Second, within this approach, interdisciplinarity is addressed as a potential means of advancing intellectual frontiers. Third, with ecosystems research often paralleling environmental issues of social and political importance, interdisciplinarity is presented as a necessary part of improving the giving of advice and environmental governance. Before pursuing each of these discussions in turn it is first necessary to ask what it means to be interdisciplinary.

Being interdisciplinary – rationales and definitions

Being interdisciplinary is not easy. Research institutes and universities are beset by multiple divisions and hierarchies. These are most apparent in divisions organised along departmental lines. Yet, even within these divisions, fractures exist between academics based on methodology, approach and subject matter. In ecology, for example, there are debates about the validity of using the results of experimental approaches in the field versus those in the laboratory or modelled on computers to address the same question (Carpenter 1996).

Specialisation, in part, has been a consequence of an enquiring culture which has stretched academic research into an increasing number of areas and pursued knowledge in increasing depth. However, intellectual boundaries and divisions are also the outcome of a culture where contestation and debate are not only the norm in academic research, but necessary elements for generating rigorous and robust knowledge. Between disciplines and within disciplines, academics not only produce knowledge, but also seek to uphold and defend those perspectives against alternatives. Knowledge is rarely singular and agreed, but varied, contradictory and political. Whether through publication, presentation or competition for funding, academic effort is held up to fierce scrutiny, argument and peer review. Indeed, the ability to ascertain and ensure research rigour is dependent on these processes. This has led Nowotny to note that if academics are divided in the production of knowledge, they are united by a culture of competition and an ethos of critical engagement (Nowotny *et al.* 2001).

It is not, therefore, surprising that such disciplinary boundaries can be deeply entrenched and are often actively maintained in the cultures and structures of universities and research institutes. For instance, while current discourse espouses the virtues of collaboration, researchers are faced with significant disincentives to working outside their own disciplines. It is often perceived to be safer to develop careers within well-defined (and by implication, monodisciplinary) research programmes (Meagher and Lyall 2005). Publishing joint papers outside of one's own disciplinary base may be detrimental to career progression so long as career progression remains largely tied to one's own discipline (Creamer 2005). Doctoral students interested in interdisciplinary study may be deterred not only by the intellectual endeavour required, but also by the perceived risks to progression and career development involved. Even within most university campuses, disciplinary divisions are spatially reinforced, with the physical sciences housed in buildings in one district of the campus, medical science in another, the social sciences and humanities in others and so on (Hall *et al.* 2006).

The divisions shaping academic endeavour are pervasive and enduring. While disciplinary divides and specialisation have been influential in the development of environmental research, it is also clear that they come with some problems. Foremost, it is charged that an enduring emphasis on disciplinary-focused scholarship has led to a conservative approach to knowledge which, as a consequence, has burdened the academy with an inability to respond to today's pressing environmental challenges. Compartmentalised and partial understandings of ecological phenomena are being produced, when multi-faceted and contextualised research is needed. For instance, climate change, sustainable development and global security each have implications for, and are being addressed by, a wide range of disciplines in the social and biological sciences. Stated differently, the footprint of environmental issues, as Daily and Ehrlich (1999) point out, oversteps the disciplinary boundaries we construct in our everyday research practices.

A first rationale for interdisciplinarity thus posits the need to work across boundaries, generating greater engagement amongst academic and research communities to understand complex environmental problems and to respond to them (Klein 1996). What is required, as Lowe and Phillipson (2006) suggest, is a counterbalance to division so as to ensure 'dynamic', 'fluid' and 'networked' systems of knowledge production. Importantly, the argument being put forward by those advocating interdisciplinary research is not simply that complex issues require multiple interpretations. Rather, further insights and opportunities are believed to exist in the relationships and linkages between disciplines and knowledges. Speaking about interdisciplinarity amongst the physical sciences, although equally as pertinent for discussion here, the UK Minister of State for Science and Innovation framed the issue in the following terms:

> Nature does not recognise or differentiate between biology, physics and chemistry. Increasingly the boundaries between the disciplines are becoming blurred and many of the most interesting scientific questions are about the interfaces and linkages between traditional subject areas (Pearson 2007).

A second rationale for collaboration is thus that working together can enable intellectual and empirical development. Interdisciplinarity, as Kinzig (2001) argues, is essential to pushing forward the frontiers of academic and environmental inquiry. Interdisciplinarity is thus not only integrative, but also potentially transformative. Further insight into these two intersecting rationales can be drawn from a brief exploration of some of the different ways in which collaboration is imagined and defined.

Multi-, inter- and trans-disciplinarity

So far in this chapter, interdisciplinarity collaboration has been discussed in general terms to describe a practice, or engagement, which combines research approaches across boundaries and which seeks to integrate knowledge around shared problems. However, within the literature on the subject it is increasingly common to develop terminologies which not only differentiate disciplinary from interdisciplinary practices, but which further distinguish a variety of collaborative practices based on the degree of synthesis supported (Aboelela *et al.* 2007). For instance, it is common to differentiate between multi-, inter- and trans-disciplinary research (see Rosenfield 1992, or Tress *et al.* 2005). Although there is no agreed set of definitions for these practices, a useful contribution to this discussion is made by Rosenfield's (1992) research on collaboration between the health and social sciences.

He defines *multi-disciplinarity* as the most rudimentary and pervasive of practices, involving researchers working on a shared problem, but autonomously, and maintaining individual disciplinary perspectives. Integration is minimal and might only include the parallel publication of research results at the end of the project. *Inter-disciplinarity* sees greater joint collaboration, with researchers working together to address overlapping research topics, but the value of each contribution is still largely understood according to its individual disciplinary basis. Finally, it is becoming more common to advocate *trans-disciplinarity* and the need for greater integration involving working together to identify research problems, frame questions and engender unique research approaches and analysis. Nowotny (2003) finds a semantic appeal in the prefix 'trans' because it points toward 'transgressing' boundaries, as opposed to simply working across them. Similarly, Rosenfield (1992) speaks of the necessity to 'transcend individual disciplinary perspectives' so as to enable the development of 'common conceptual frameworks'. Trans-disciplinary research seeks to reshape the academy and to generate novel approaches to theory and research.

Within both multi- and inter-disciplinary research we can identify instrumental rationales for applying collaboration as a means of addressing shared problems. However, little acknowledgement is made of the potential limitations of these engagements. There remains a lack of reflexivity and innovation in the way in which knowledge is produced or applied. Trans-disciplinarity sets out loftier ambitions for collaboration. As Klein (1996) argues, transgressing the divisions between current disciplines involves drawing new boundaries and identifying new knowledge spaces and institutional frameworks. Scoones (2004) similarly asserts the need for greater collaboration between social scientists interested in perceptions of the environment and the ecologists who study and quantify the natural interactions within the environment. From the position of the social scientist, he argues that developments in 'new ecology' which address complexity, uncertainty and the dynamics of ecological processes have implications for how social scientists conceptualise and research social–natural relations. Collaboration is advanced as a means of opening disciplinary approaches and practices to *reflexive* consideration, development and change (Lowe and Phillipson 2006, Lattuca and Creamer 2005, Kinzig 2001). In other words, it is in trans-disciplinary research that the collaborative study of the environment – aimed at advancing intellectual frontiers – is articulated. Below, we turn to addressing how collaboration understood in these terms can contribute to, and advance, ecosystems and ecological research.

Science and society: developing social robustness in ecosystems research

While there has been a traditional tendency to treat humanity as exogenous to ecosystems, it is today difficult to find any situation in which ecosystems are not impacted on by human social relations. Social scientific concerns with demographics, economics and development overlap the same intellectual spaces that consider the dynamic functioning of the environment. Similarly, ecosystems research is increasingly linked to attempts to manage or govern human activities in order to protect the structure and function of ecosystems. Directing and applying ecosystems research to address environmental problems require scientific understanding, but also entail the recognition of science and ecology as social actors, operating within a social milieu (Ewel 2005). There is, thus, a pressing need to recognise the social as well as the ecological nature of ecosystems in order to cope with the complexity of social–natural relations and to identify fruitful means of environmental governance.

Confronting the social in the natural compels a critical reflection of the ways in which research is developed in relation to ecology and the environment. Nowotny *et al.* (2001) describe this impetus as the necessity of shifting towards more 'socially robust' forms of science (also Stilgoe *et al.* 2006). In contrast to traditional assumptions that have identified science as autonomous from society, Nowotny *et al.* (2001) assert the need to situate scientific practice within a social context. Foremost, this means coming to terms with science as a social process through which knowledge is actively produced and utilised, as opposed to simply revealed and applied straightforwardly as evidence (Barnes and Edge 1982, Latour and Woolgar 1986, Knorr-Cetina 1999).

We will return to this subject below, but for now the implication of socially contextualising science is that science is exposed to greater social scrutiny.

Instead of autonomous scientific facts demanding absolute authority, placing science within a social context exposes its uncertainties and partialities. Indeed, instead of conferring automatic trust, scientific evidence is as likely to engender social misgivings or ambivalence. This scientific malaise is something which sociologists, such as Ulrich Beck (1992) and Anthony Giddens (1991), describe as central to the consciousness of contemporary western society. The techno-scientific progress of the last two centuries is now recognised to have come with significant costs. Likewise, uncertainty in the ability to understand and ameliorate risk is plainly apparent. Environmental issues are prominent in this context. Many of the problems we face as a society are only now becoming apparent and our understanding of these problems is most often partial, or speculative. Even in contexts where knowledge of a problem is stable and agreed, the way forward is opaque at best. The environmental sciences find themselves in a highly politicised and contentious social space where recourse to discourses of absolute and autonomous facts is both naive and inappropriate.

Social robustness, then, is the capacity to develop and apply knowledge which has the ability to negotiate and contextualise uncertainty and the wider processes of social accountability. Its development is dependent on the reflexive development and the transformative capacity associated with transgressing disciplinary boundaries. Calls for interdisciplinarity thus parallel wider epistemological challenges to science in general, and to environmental science in particular. The remainder of this chapter turns towards a discussion of where social robustness may be suggested to contribute to the development of ecological science and its role in environmental governance.

Contextualising ecology

One area where greater interdisciplinary engagement may be of benefit relates to the ability to understand ecology as a social as well as a scientific process. In part, this involves the active and reflexive understanding of the contribution, and the limitations, of ecological research to issues of social environmental concern. Moreover, it involves addressing ecology and environmental knowledge as the products of social contexts, as well as being involved in their production.

Identifying ecologists as productive actors, and by implication involved in the shaping of social knowledge, is controversial. It is more common to think of nature and the environment as fixed and the role of ecological enquiry as a process of revelation as opposed to construction. Indeed, constructive theories of science are frequently responded to with concerns about eroding the authority of scientific knowledge and exposing environmental knowledge and governance to epistemological relativism. Yet, while the constructivist perspective does pose challenges to science in this regard, the nature of this criticism is somewhat less dramatic. The suggestion is not that ecological phenomena or environmental degradation are any less real, or that all knowledges are equal, but rather that ecological knowledge can only be arrived at, described, assessed and applied through human subjectivities, and by implication through social relations. Scientific knowledge is inevitably socially mediated. Through cumulative empirical enterprise, scientists are both responsive to social contexts, and also engaged in their construction.

Consider our knowledge of the natural world, which is neither always stable, nor often agreed upon. It is mediated by our empirical and social interactions with nature. Indeed the term 'environment' is probably more appropriate here, as it signifies an understanding of the natural world which is inherently viewed from the multiple perspectives of human experience and habitation. The instability of human perspectives of the environment is revealed in the historical movement in how the environment is understood, and how these understandings shape human action. Our notion of the environment is continually shifting from perceptions of mastery and control associated with enlightenment and industrialisation (see Leiss 1994, Harvey 1996), through the early recognition of environmental degradation and the development of conservation movements (Leopold 1968, Carsons 1963), to today's emphasis on the global and uncertain nature of risk and the implications of environmental degradation for human well-being (Lash *et al.* 1996).

In practice, we can see ecology as socially mediated. A look across the discipline makes it difficult to define what ecology, ecosystems research and their governing principles are. Rather they are fields characterised by complexity and are openly debated and contested (for example, Cherrett 1989). All ecological research involves choices about research priorities and how the subject is approached, which conversely involve other choices about what is less important, and what approaches are less suitable. For instance, at what scale(s) does the research address the ecosystem? Should research focus on a specific experimental microcosm or on whole systems (Carpenter 1996)? Empirical vision and analytic modelling are always partial. As Macfadyen (1975) noted in his presidential address to the British Ecological Society in 1973, while ecologists can be encouraged to follow interactions and ecological relations, an understanding of the totality is likely to elude them. Kokko likens it to a map of the countryside:

> Maps are models that are designed to help us grasp certain features of the landscape. For example, a map consists of contour lines which help us predict which way a river will flow once we stumble across it. But a map would become completely useless if it had every tuft of grass marked on it... [S]taring at a too detailed model teaches us nothing more than staring at the original ecosystem with its complete mess of evolutionary and ecological detail (Kokko 2005, p. 1155).

Similarly, ecologists seeking not only to identify and understand ecosystems but to protect them from human activities or restore them face difficult normative dilemmas. Questions such as – What is the best conduct for restoring an ecosystem? To what status should it be restored? For what purposes is it being restored? – are not empirical, but social questions (Higgs 2005, Turner 2005), the solutions to which, Higgs suggests:

> will not come from regression analyses, or replicated studies, but the deep, searching, intelligent, humble inquiries into the human past and prospect, to the varieties of human experience, value and creativity, and of course to the many ways we have both loved and despoiled nature (Higgs 2005).

The contextualisation of ecological research is of particular consequence when considering the relationship between research, expert advice and governmental policy. Empirical choices can translate into advice about research priorities – which species, which interactions within the ecosystem, and at which level should governments be addressing environmental problems? It is also necessary to question how empirical choices lead to government priorities in generating regulation, protection or conservation.

These questions can be explored further with reference to the evolving role of ecosystems research in marine governance. Marine systems cover more that 70 per cent of the Earth's surface and the goods and services which these systems provide have been valued at $\$20,949 \times 10^9\ yr^{-1}$ (Costanza *et al.* 1997). Despite humans' largely land-locked existence and the importance of marine systems to human well-being, a recent study has shown that the impacts of human activities could be detected in all 20 of the marine ecosystems examined in the model and 41 per cent of these areas were considered to be strongly impacted by multiple factors related to human activities (Halpern *et al.* 2008).

There is a progressive recognition of this point in the management of marine systems and of the knowledge which successful governance requires. Whilst traditional approaches have focused on generating data documenting resource depletion – e.g. the effects of fishing on commercial fish stocks – attempts are now being made to initiate more holistic approaches which formally integrate wider environmental and social management issues. For instance, environmental economics is increasingly in demand from national governments and is being used to assess the impact of various management scenarios on nature in monetary terms (Moran *et al.* 2008). While putting a monetary value on nature is a concept which is uncomfortable for many ecologists (McCauley 2006), environmental economics brings the socio-political issues at the heart of fisheries management into explicit focus. Instead of addressing declining fish stocks in isolation, fisheries resources can be understood in relation to employment, their market value, and more importantly, in relation to the loss of economic value associated with a damaged marine ecosystem. Understanding and managing marine ecosystems involve a wide range of social and social–natural interactions, and consequently an equally diverse range of social networks and actors (for example, Gray and Hatchard 2008).

Government legislation and advice are now beginning to recognise the interdisciplinary nature of management and in some cases explicitly require that ecosystem 'functions' and 'goods and services' are protected (EC 2000, EC 2007a). The Millennium Ecosystem Assessment (MA) and the Convention on Biological Diversity (CBD) now link ecosystems with human well-being and sustainable development, as discussed in Frid and Raffaelli (this volume). This new approach explicitly recognises humankind as a component of the ecosystem and requires greater integration between industry, conservation groups, scientists and other stakeholder groups to protect fish stocks and the ecosystem which supports them. It contrasts with traditional academic definitions, which identify ecosystems as purely bio-physical constructs.

Ecology and the public

Placing science within a social context has further consequences for the relationship between ecology and the publics who occupy the social spaces in which ecological practice operates. Here the use of 'publics', instead of 'public', is a conscious one. Ecologists involved in research linked to environmental governance will find themselves exposed to a wide variety of competing public interests. These include stakeholders from industry, local communities, environmental organisations, governmental policy advisers and so forth. These complex political relationships can be challenging; a situation which is further exacerbated by the often contentious and uncertain relationship between science and the public.

The traditional, and often still prevailing, attitude towards this relationship amongst scientists and policy makers remains rooted in the belief that scientific expertise can be viewed apart from, and above, public politics. Where controversy has broken out around environmental science, the tendency has been to generate a public understanding of science, and to counter public ignorance through programmes of education and engagement (Irwin 1995, Irwin and Wynne 2004). The Royal Society (2004), for instance, identified the public understanding of science as a priority, arguing that it was the 'duty' of scientists to get out, communicate and explain science to the public. Similarly, an emphasis on evidence-based policy making has legitimated technocratic approaches to governance where public concerns and debates about risk have been ostracised by an emphasis on scientific risk assessment (Jasanoff 1990).

This approach to citizen–science relations has been heavily criticised in the social sciences, as well as by parts of the wider polity (for example Irwin and Wynne 2004, Nowotny *et al.* 2001, Wilsdon and Willis 2004). The approach is perceived as problematic as it imposes an asymmetric, hierarchical and paternalistic relationship between experts and members of the public. As Fuller

(2000) notes, the public is expected to trust scientists, but there is little thought to making science accountable to the public. Lord Phillips (2000), in his report from the enquiry into the government mishandling of Bovine Spongiform Encephalopathy (BSE), or 'mad cow disease', in Britain, thus identifies public trust in science and scientific governance as having been further casualties of the disease. Consistent assumptions about public irrationality and ignorance, combined with concerns about protecting the British beef industry, saw public concerns dismissed and messages of scientific reassurance given, even when the empirical basis of that advice was itself being questioned (Jones 2005).

Thus, a central aspect of social robustness in science involves addressing this problematic relationship. Fundamentally, this requires the awareness that if science is recognised to be contextual, uncertain and partial, then the knowledge and expertise it produces must be accepted as conditional (Nowotny *et al.* 2001, Royal Society 2004). Instead of science being perceived as absolute and applied to tidy up complex environmental policy areas, governments are now challenged to explore complex environmental issues fully and to contend with routine uncertainty. Partial scientific expertise has to be reconciled with a range of social questions about the nature of risk in environmental governance. What risks are acceptable given the benefits they offer, and which are not? What risks can be managed, and which cannot? What degree of precaution should be taken in developing policy and regulation when risks are either unknown, or fully understood? The environmental sciences have an essential role to play in helping governments' response to these questions, but cannot answer them alone. As such, science is increasingly challenged to become more accountable to publics, and participate in processes of governance which employ novel means of integrating a wider range of expertise, knowledge and experience.

Arguments for a more socially robust science overlap an infusion of deliberative ideas about governance and participation, of which the opening up of science to scrutiny and debate is a key aspect (Hagendijk and Irwin 2006). Scientists are challenged to move beyond the deficit assumptions associated with traditional approaches to the public understanding of science, to embrace more open and dialogical relations. Instead of communicating science to the public, there has been a more recent move to create engagements and dialogue where the public can communicate their concerns, thoughts and experiences to scientists and governments.

The importance of engagement is partly normative, but the possibility of the public making important contributions to scientific knowledge of the environment, as well as to environmental governance has also been raised. In other words, does engagement offer scientists or government cognitive gains? What can local communities tell scientists about the ecosystems in which they live and work? How might publics be involved in environmental monitoring and management? The public within this perspective are not seen as lacking knowledge, but holding important experiential knowledge which could be of use to the ecologist and the policy maker (see also, White *et al.*, this volume).

Alongside these instrumental contributions, it is also suggested that publics, through engagement, can play important roles in encouraging social reflexivity in science. For instance, where it can be recognised that scientific knowledge is incomplete and limited, engagement could be a means of drawing these limitations out, as well as considering what other forms of knowledge might be beneficial in filling these gaps. Engagement can be a means of encouraging scientists to relate their research to a wider social context, but in doing so engagement will expose science to difficult questions about its uncertainties and its limitations in applicability. When scientists enter into dialogue with the public or different publics, the boundaries which frame the subject discussed are often pulled apart. Dialogue and engagement lead to a loosening of control over scientific discourses. In a recent project on the inclusion of lay membership on scientific advisory committees (Jones *et al.* 2008, Jones and Irwin 2009), the ability of non-scientists to encourage openness in scientific debates and to engage in critical and rigorous debates about uncertainty were referred to as 'challenge functions'. While the loosening of control of ecological issues can be difficult and frustrating, it remains essential to processes of good environmental governance. The rights of stakeholders to participate in decisions pertaining to the environment have been enshrined in the Århus Convention (1998) which addresses the need to develop formal mechanisms to deal with the conflicting interests of stakeholder groups in the management of natural resources.

In marine systems, initiatives to 're-engage' publics in research have advanced rapidly in recent years. The failure of traditional single stock fisheries management in Europe led to the reform of the EU Common Fisheries Policy in order to develop more holistic ecosystem-based approaches (EC 2002). Where traditional management strategies were largely top-down, and often supported and justified by narrowly focused and fragmented ecological science, ecosystem-based approaches explicitly recognised humankind as a component of the ecosystem. This has necessarily required greater integration between industry, conservation groups, scientists and other stakeholder groups to protect fish stocks and the ecosystem which supports them.

Gray and Hatchard (2008) assessed the success of twenty-two EC-funded fisheries projects in integrating stakeholder participation with the scientific approaches which underpin ecosystem-based fisheries management (EBFM). Their conclusions suggest that EBFM will significantly extend the 'reach of democratic culture' in the marine environment and that environmental stewardship will become increasingly more prominent in fisheries governance as a means of empowering stakeholders. Partly, this is a consequence of a perceived ethical duty to include publics in a context in which ecological research has a profound impact on environmental protection and management, as well as on the communities and publics involved in the fisheries. Moreover, they note the successes and failures of management, and remediation will probably be dependent on scientists working together with local communities. Local communities, for instance, can hold valuable experiential knowledge about the ecosystems they interact with, or be positively involved in monitoring and management activities. Likewise, it is possible to speculate that interaction with publics will be essential in encouraging attitudes and behavioural changes towards marine ecosystems, both in local communities and amongst a wider public of consumers.

The need for greater interdisciplinary collaboration is only likely to increase with the development of marine spatial planning initiatives, such as those being drafted for the European Maritime Policy (EC 2007b). Using the experience of spatial management tools, such as marine protected areas, which restrict human activities in specific sites and which may thereby impact on certain communities or industries disproportionately, the demand for robust evidence to support the social, ecological and economic benefits that these tools may deliver is extremely high and politically charged. Scientific understanding of the dynamics of ecosystems and how they function is extremely limited due to their immense complexity (Frid *et al.* 2006), and the development of governance structures which can robustly address the range of conflicts which are likely to occur with marine spatial planning is urgently required.

Conclusions

In C.P. Snow's 1959 Rede Lecture (1998) entitled the 'Two Cultures', he raised a concern about the need to enhance the role of science in education and society. In particular, he looked to the lack of general scientific awareness and knowledge across the university campus in the other disciplines. In today's academic lexicon, the 'Two Cultures' is often read as a statement of the incontrovertible differences between the 'hard' and 'soft' sciences. However, while antagonisms may persist along these lines, Snow's intention was not to divide academic study, but to draw links across these barriers. He was not calling for the demise of non-scientific approaches within the university, but for an end to their antagonism and ignorance with the developing sciences. Indeed, the examples given in this paper strongly support not only the need for a mature and confident ecological science, but also for the crucial importance of ecological research in addressing deeply worrying environmental contexts in which we all share an interest. However, Snow failed to ask the question of what science may learn from outside its own boundaries.

In this chapter, we sought to elucidate the many possibilities of interdisciplinarity between the environmental sciences, ecology in particular, and the social sciences. It is clear that the environmental challenge does not respect socially erected and maintained disciplinary boundaries. Our ability to understand the shared problems we face as a society, and our ability to respond to them, require a change in approach. There is an urgent need to promote interdisciplinarity to provide engagement in knowledge production and 'joined-up' advice to managers and policy makers and to develop tools which can address complex management issues.

Collaboration between the sciences and the social sciences may be part of the means to generate greater interdisciplinarity around environmental issues. Sharing information, dialogue, and participating together on shared research grants – now an increasing requirement of many funding bodies – may all be beneficial. Delineated as *multi-* and *inter*-disciplinarity by Rosenfield (1992), it is in these practices where the most substantial developments can be seen. However, we have also posited an argument for the potential of more substantial *trans*-disciplinary shift. The authors' observations of working in multi- and inter-disciplinary teams in marine systems has shown that in practice it is very difficult to make the next step into trans-disciplinarity as the language, expectations and academic theories of the different disciplines are often either not understood or misunderstood. It takes a significant amount of time and effort to develop trans-disciplinary approaches and in many of the larger multi-institute projects this is difficult to achieve and the difficulties are exacerbated in larger projects where the partners are often in different countries and the effort for collaboration is limited. In reality, many trans-disciplinary projects use multi- or at most inter-disciplinary teams because of these difficulties, but 'inter-disciplinarity has not reached its escape velocity' (John Lawton, quoted in Meagher and Lyall 2005). Trans-disciplinarity cannot be achieved through discrete series of shared workshops, meetings or seminars. Instead it require a commitment to reflexively engaging with our own epistemologies and research frameworks. To achieve this requires openness to alternative ways of knowing and the evolution of our own perspectives of research, expertise and education.

References

Århus Convention (1998) Convention on access to information, public participation in decision-making and access to justice in environmental matters. *Europa* IP/04/1516.

Aboelela, S.W., Larson, E., Bakken, S., Carrasquillo, O., Formicola, A., Glied, S.A., Haas, J. and Gebbie, K.M. (2007) Defining interdisciplinary research: conclusions from a critical review of the literature. *Health Services Research* **42**(1), 329–46.

Barnes, B. and Edge, D. (1982) *Science in Context: Readings in the Sociology of Science*. Milton Keynes, Open University Press.

Beck, U. (1992) *Risk Society: Towards a New Modernity*. London, Sage.

Carpenter, S.R. (1996) Microcosm experiments have limited relevance for community and ecosystem ecology. *Ecology* **77**(3), 677–80.

Carsons, R. (1963) *Silent Spring*. London, Hamilton.

Cherrett, J.M. (1989) Key concepts: the results of a survey of our members' opinions, pp. 1–16. In J.M. Cherrett (ed.) *Ecological Concepts: The Contribution of Ecology to an Understanding of the Natural World*. Oxford, Blackwell Scientific Publications.

Costanza, R., d'Arge R., de Groot, R., Farber, S., Grasso, M., Hannon, B., Limburg, K., Naeem, S., O'Neill, R.V., Paruelo, J., Raskin, R.G., Sutton, P. and van den Belt, M. (1997) The value of the world's ecosystem services and natural capital. *Nature* **387**, 253–60.

Creamer, E.G. (2005) Promoting the effective evaluation of collaboratively produced scholarship: a call to action. *New Directions for Teaching and Learning* **102**, 85–98.

Daily, G.C. and Ehrlich, P.R. (1999) Managing Earth's ecosystems: an interdisciplinary challenge. *Ecosystems* **2**(4), 277–80.

EC (2000) Directive 2000/60/EC of the European Parliament and of the Council of 23 October 2000 establishing a framework for Community action in the field of water policy.

EC (2002) Council Regulation no. 2371/2002 of 20 December 2002 on the conservation and sustainable exploitation of fisheries resources under the Common Fisheries Policy.

EC (2007a) Recommendation for Second Reading on the Council commons position for adopting a directive of the European Parliament and of the Council establishing a Framework for Community Action in the field of Marine Environmental Policy (Marine Strategy Framework Directive) (9388/2/2007 – C6-0261/2007 – 2005/0211(COD)).

EC (2007b) Communication from the Commission to the European Parliament, the Council, the European Economic and Social Committee and the Committee of the Regions on an Integrated Maritime Policy for the European Union. Brussels, 10.10.2007. COM(2007) 575 final.

Ewel, K. (2005) Natural resource management: the need for interdisciplinary collaboration. *Ecosystems* **4**, 716–22.

Flyvbjerg, B. (2001) *Making Social Science Matter: Why Social Inquiry Fails and How it Can Succeed Again*. Cambridge, Cambridge University Press.

Frid, C.L.J., Paramor, O.A.L. and Scott, C.L. (2006) Ecosystem-based management of fisheries: is science limiting? *ICES Journal of Marine Science* **63**, 1567–72.

Fuller, S. (2000) *GOVERNANCE OF SCIENCE: Ideology and the Future of the Open Society (Issues in Society)*. Milton Keynes, Open University Press.

Giddens, A. (1991) *The Consequences of Modernity*. Cambridge, Polity Press.

Gray, T. and Hatchard, J. (2008) A complicated relationship: Stakeholder participation and the ecosystem-based approach to fisheries management. *Marine Policy* **32**(2), 158–68.

Hagendijk, R. and Irwin, A. (2006) Public deliberation and governance: engaging with science and technology in contemporary Europe. *Minerva* **44**(2), 167–84.

Hall, J.G., Bainbridge, L., Buchan, A., Cribb, A., Drummond, J., Carlton, G., Hicks, T.P., Mcwilliam, C., Paterson, B., Ratner, P.A., Skarakis-Doyle, E. and Solomon, P. (2006) A meeting of minds: interdisciplinary research in the health sciences in Canada. *Canadian Medical Association Journal* **175**, 763–71.

Halpern, B.S., Walbridge, S., Selkoe, K.A., Kappel, C.V., Micheli, F., D'Agrosa, C., Bruno, J.F., Casey, K S., Ebert, C., Fox, H.E., Fujita, R., Heinemann, D., Lenihan, H.S., Madin, E.M.P., Perry, M.T., Selig, E.R., Spalding, M., Steneck, R. and Watson, R. (2008) A global map of human impact on marine ecosystems. *Science* **319**, 948–52.

Harvey, D. (1996) *Justice, Nature and the Geography of Difference*. Oxford, Blackwell.

Higgs, E. (2005) The two-culture problem: ecological restoration and the integration of knowledge. *Restoration Ecology* **13**(1), 159–64.

Huber, L. (1992) Towards a New Studium General: some conclusions. *European Journal of Education* **27**, 285–301.

Irwin, A. (1995) *Citizen Science: A Study of People, Expertise and Sustainable Development.* London, Routledge.

Irwin, A. and Wynne, B. (eds.) (2004) *Misunderstanding Science?* Cambridge, Cambridge University Press.

Jasanoff, S. (1990) *The fifth branch: science advisers as policymakers.* Cambridge, MA, Harvard University Press.

Jones, K.E. (2005) *Understanding Risk in Everyday Policy Making.* Defra, London.

Jones, K.E. and Irwin, A. (2009) Creating space for engagement: lay membership in contemporary risk governance. In B. Hutter (ed.) *Anticipating Risks and Organizing Risk Regulation in 21st Century.* Cambridge, Cambridge University Press.

Jones, K.E., Irwin, A., Farrelly, M. and Stilgoe, J. (2008) *Understanding Lay Membership and Scientific Governance.* London, Defra.

Kinzig, A.P. (2001) Bridging disciplinary divides to address environmental and intellectual challenges. *Ecosystems* **4**, 709–15.

Klein, J.T. (1996) *Crossing Boundaries: Knowledge, Disciplinarities and Interdisciplinarities.* London, University Press of Virginia.

Knorr-Cetina, K. (1999) *Epistemic Cultures: How the Sciences Make Knowledge.* Harvard University Press, Cambridge, MA.

Kokko, H. (2005) Useful ways of being wrong. *Journal of Evolutionary Biology* **18**, 1155–7.

Lash, S., Szerszynsky, B. and Wynne, B. (1996) *Risk, Environment and Modernity: Towards a New Ecology.* London, Sage.

Latour, B. and Woolgar, S. (1986) *Laboratory Life: The Construction of Scientific Facts.* Princeton University Press, New Jersey, USA.

Lattuca, L.R. and Creamer, E.G. (2005) Learning as professional practice. *New Directions for Teaching and Learning* **102**, 3–11.

Leiss, W. (1994) *The Domination of Nature.* Montreal, McGill-Queen's University Press.

Leopold, A. (1968) *A Sand County Almanac and Sketches Here and There.* London, Oxford University Press.

Lord Phillips (2000) *The BSE inquiry Vol 1: findings and conclusions.* London, The Stationery Office, HC 887–1.

Lowe, P. and Phillipson, J. (2006) Reflexive interdisciplinary research: the making of the rural economy and land use. *Journal of Agricultural Economics* **57**(2), 165–84.

Macfadyen, A. (1975) Some thoughts on the behaviour of ecologists. *Journal of Animal Ecology* **44**(2), 351–63.

McCauley, D.J. (2006) Selling out on nature. *Nature* **443**, 27–8.

Meagher, L. and Lyall, C. (2005) Evaluation of the ESRC/NERC Interdisciplinary research studentship scheme. www.esrcsocietytoday.ac.uk/ESRCInfoCentre/Images/ESRC-NERC%20Scheme%20Review%20Final%20Report_tcm6-17593.pdf.

Moran, D., Hussain, S., Fofana, A., Paramor, O.A.L, Robinson, L.A., Winrow-Giffin, A. and Frid C.L.J. (2008) *The Marine Bill – Marine Nature Conservation Proposals – Valuing the Benefits.* CRO380: Defra Natural Environment Group, Science Division, London.

Naeem S. (2006) Expanding scales in biodiversity-based research: challenges and solutions for marine systems. *Marine Ecology Progress Series* **311**, 273–83.

Nowotny, H. (2003) The potential of transdisciplinarity. Series. www.interdisciplines.org/interdisciplinarity/papers/5. (accessed April 2008).

Nowotny, H., Scott, P. and Gibbons, M. (2001) *Re-thinking science: knowledge and the public.* Cambridge, Polity Press.

Pearson, I. (2007) *First Sir Gareth Roberts Science Policy Lecture.* London, The Science Council.

Pickett, S.T.A., Burch, W.R. and Grove, J.M. (1999) Interdisciplinary research: maintaining the constructive impulse in a culture of criticism. *Ecosystems* **2**(4), 302–7.

Rosenfield, P.R. (1992) The potential of transdisciplinary research for sustaining and extending linkages between health and social sciences. *Social Science and Medicine* **35**, 1343–57.

Royal Society (2004) *Science in Society*. London, The Royal Society.

Scoones, I. (2004) New ecology and the social sciences: what prospects for a fruitful engagement? *Annual Review of Anthropology* **28**, 479–507.

Snow, C.P. (1998) *The Two Cultures*. Cambridge, Cambridge University Press.

Stilgoe, J., Irwin, A. and Jones, K.E. (2006) *The Received Wisdom: Opening Up Expert Advice*. London, Demos.

Tress, B., Tress, G. and Fry, G. (2005) Integrative studies on rural landscapes: policy expectations and research practice. *Landscape and Urban Planning* **70**, 177–91.

Turner, R.E. (2005) On the cusp of restoration: science and society. *Restoration Ecology* **13**(1), 165–73.

Wilsdon, J. and Willis, R. (2004) *See-through Science: Why Public Engagement Needs to Move Upstream*. Demos.

CHAPTER SIX

The links between biodiversity, ecosystem services and human well-being

ROY HAINES-YOUNG and MARION POTSCHIN
Centre for Environmental Management, School of Geography, University of Nottingham

The degradation of ecosystem services poses a significant barrier to the achievement of the Millennium Development Goals and the MDG targets for 2015.
Millennium Ecosystem Assessment, 2005, p. 18

Introduction: managing ecosystems for people

No matter who we are, or where we live, our well-being depends on the way ecosystems work. Most obviously, ecosystems can provide us with material things that are essential for our daily lives, such as food, wood, wool and medicines. Although the other types of benefit we get from ecosystems are easily overlooked, they also play an important role in regulating the environments in which we live. They can help ensure the flow of clean water and protect us from flooding or other hazards like soil erosion, land-slips and tsunamis. They can contribute to our spiritual well-being, through their cultural or religious significance or the opportunities they provide for recreation and the enjoyment of nature.

In this chapter, we will look at the goods and services that ecosystems can provide and the role that biodiversity may play in producing them,[1] specifically the contribution that biodiversity makes to people's livelihoods, to their security and to their health. In other words, we will concentrate mainly on the *utilitarian* value of biodiversity. We will also explore how these ideas link up with those of the Ecosystem Approach to environmental management and policy, and some of the implications of this for how sustainable development is defined. This does not mean that traditional ideas about the need for conservation are unimportant, rather that those making the case for biodiversity need to set these issues in a broader context, and consider whether nature has utilitarian as well as intrinsic values (see for example, Chan *et al.* 2007).

[1] While many commentators use the terms 'goods' and 'services' to distinguish between the more tangible and intangible outputs from ecosystems, others use them as synonyms. In this text we make no distinction between them and use the term 'services' to cover both.

Ecosystem Ecology: A New Synthesis, eds. David G. Raffaelli and Christopher L. J. Frid. Published by Cambridge University Press.

Ecosystems services and the Ecosystem Approach

The current interest in ecosystem services has come from several sources. The most widely acknowledged is perhaps the Millennium Ecosystem Assessment (MA 2005), which was the first comprehensive global assessment of the implications of ecosystem change for people. It came about as the result of a call in 2000 by the then UN Secretary-General Kofi Annan, to 'assess the consequences of ecosystem change for human well-being and the scientific basis for action needed to enhance the conservation and sustainable use of those systems and their contribution to human well-being'.[2] The work began in 2001 and involved over 1,300 international experts. It resulted in a series of publications in 2005 that described the condition and trends of the world's major ecosystems and the services they provide, and the options available to restore, conserve or enhance their sustainable use.

The key finding of the MA was that currently 60 per cent of the ecosystem services evaluated are being degraded or used unsustainably, with major implications for development, poverty alleviation, and the strategies needed by societies to cope with, and adapt to, long-term environmental change. The key implication, flagged up in our opening quote, was that given such trends it is unlikely that the global community would achieve the so-called Millennium Development Goals that it had set itself in 2000.[3] The elimination of extreme poverty is a key international challenge, for as the Brundtland Report[4] argued in 1987, it is one of the major factors leading to environmental degradation and loss of biodiversity. The impacts of biodiversity loss on well-being are uneven across communities, affecting those who depend most on environmental resources, such as subsistence farmers and the rural poor (Díaz *et al.* 2006). A summary of the kinds of pattern we now see emerging is to be found in the first report of the study initiated by the G8+5 meeting in March 2007, on *The Economics of Ecosystems and Biodiversity* (European Commission 2008).

Although important, the Millennium Ecosystem Assessment is not the only stimulus to the current interest in ecosystem services. In fact, the idea has a longer history. Following Mooney and Ehrlich (1997), Cork *et al.* (2001) trace the development of the concept to the 1970 Study of Critical Environmental Problems (SCEP 1970), which first used the term 'environmental services'. It is possible that elements of the idea can be found even earlier, in Leopold's *Sand County Almanac* (Grumbine 1998). Nevertheless, Holdren and Ehrlich (1974) went on to refine the list of services proposed in the SCEP study, referring to them as 'public service functions of the global environment'. Westman (1977) later reduced this to 'nature's services' and finally the term 'ecosystem services' was

[2] www.millenniumassessment.org/en/About.aspx (accessed 24th July, 2008)
[3] www.un.org/millenniumgoals/ (accessed 24th July, 2008)
[4] www.worldinbalance.net/pdf/1987-brundtland.pdf (accessed 24th July, 2008)

used by Ehrlich and others in the early 1980s (Mooney and Ehrlich 1997). The concept is also specifically covered by the principles underlying the Ecosystem Approach as set out in the Convention for Biological Diversity (CBD).[5]

As described elsewhere in this volume, the Ecosystem Approach emerged as a topic of discussion in the late 1980s and early 1990s amongst the research and policy communities concerned with the management of biodiversity and natural resources (Frid and Raffaelli, this volume; see also Hartje *et al.* 2003). A new focus was required to achieve robust and sustainable management and policy outcomes and an Ecosystem Approach, it was suggested, would deliver more *integrated* policy and management at a landscape scale and be more firmly directed towards human well-being.

According to the CBD, the Ecosystem Approach seeks to put human needs at the centre of biodiversity management. If we are to ensure that decisions take full account of the value of natural resources and biodiversity, then the links between biodiversity and well-being must be clear – hence the emphasis that the Convention places on identifying the benefits from nature. Under the Convention, the Ecosystem Approach forms the basis for considering all the services provided to people by biodiversity and ecosystems in a holistic framework (Secretariat of the Convention on Biological Diversity 2004).

The design of environmental management strategies or policies for future development often involves weighing up the consequences of proposed actions. We need to consider impacts upon ecosystems as well as the social and economic systems to which they are linked so that the choices society makes are as well informed as possible (Potschin and Haines-Young 2006). Thus questions about what kinds of service an ecosystem can provide, how much of these services we need now and in the future, and what might threaten their output are fundamental. Ecologists have much to contribute to such debates. Decisions about policy and management may ultimately be a matter of societal choice but as the Ecosystem Approach recognises, those decisions have to be grounded in a good understanding of the biophysical limits that constrain ecological processes and the spatial and temporal scales at which they operate. Before we can take the Ecosystem Approach forward, we need to explore the science that underpins these ideas.

Ecosystem service typologies

Although we can define an ecosystem service in fairly simple terms, as 'the benefits ecosystems provide' (MA 2005, p.1), difficulties can arise when applying the concept in an operational setting. A number of typologies (categorisation of different types of service) have been proposed. In the typology suggested by the MA, four broad types of service were recognised, namely: those that

[5] www.cbd.int/ecosystem/principles.shtml (accessed 24th July, 2008)

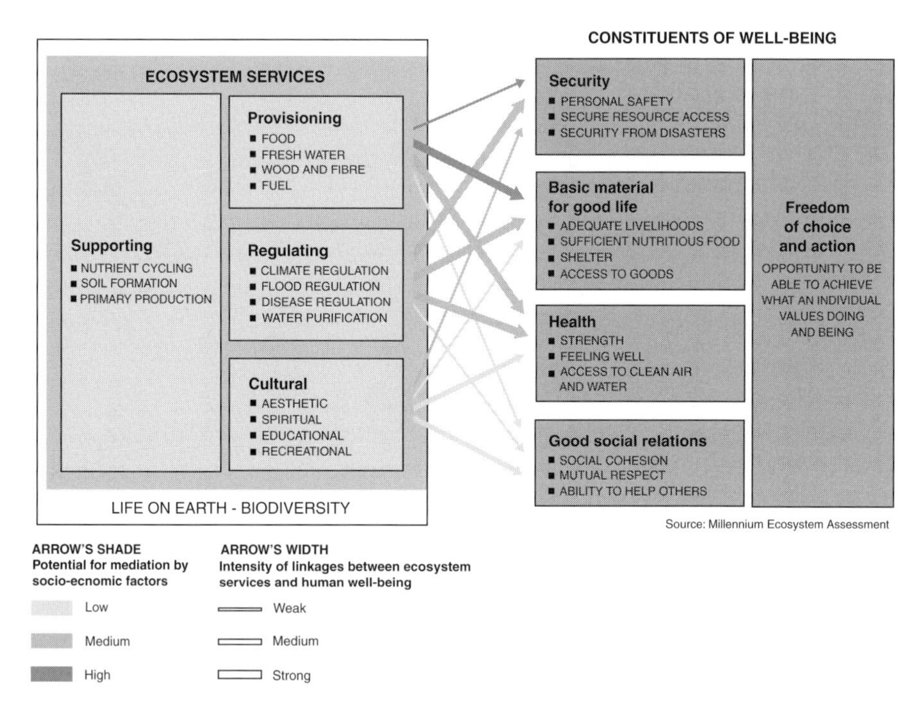

Figure 6.1 The links between ecosystem services and human well-being (after MA 2005).

cover the material or *provisioning* services; those that cover the way ecosystems *regulate* other environmental media or processes; those related to the *cultural* or spiritual needs of people; and finally the *supporting* services that underpin these other three types. Examples of services under each of these broad headings, and their relationship to different components of human well-being, are illustrated in Figure 6.1 and Table 6.1. The typology shown in Table 6.1 is from Kremen (2005), but it is based on the MA. It is particularly useful because it also attempts to detail some of the ecological and spatial characteristics of the services.

It is important to note features of the typology and relationships shown in Figure 6.1. First, 'biodiversity' *per se* is not a service; rather, the MA represents services as flowing directly from the presence of life on earth. This is an important point, because it suggests that ecosystem services depend fundamentally on the structures and processes generated by living organisms and their interactions with, and processing of, abiotic materials. As a result some commentators (Swallow *et al.* 2007, Smith 2006) think it may be useful to distinguish between ecosystem services that are a consequence of biodiversity, and a more general class of 'environmental services', like wind or hydraulic potential, that have a more indirect connection. Wind or hydraulic flows may be affected by the presence of living organisms, but ecological processes are not primarily responsible for them.

Table 6.1. *A typology of ecosystem services and their ecological characteristics (after Kremen 2005).*

Service	Ecosystem service providers/ trophic level	Functional units	Spatial scale	Potential to apply this conceptual framework for ecological study
Aesthetic, cultural	All biodiversity	Populations, species, communities, ecosystems	Local–global	Low
Ecosystem goods	Diverse species	Populations, species, communities, ecosystems	Local–global	Medium
UV protection	Biogeochemical cycles, micro-organisms, plants	Biogeochemical cycles, functional groups	Global	Low
Purification of air	Micro-organisms, plants	Biogeochemical cycles, populations, species, functional groups	Regional–global	Medium (plants)
Flood mitigation	Vegetation	Communities, habitats	Local–regional	Medium
Drought mitigation	Vegetation	Communities, habitats	Local–regional	Medium
Climate stability	Vegetation	Communities, habitats	Local–global	Medium
Pollination	Insects, birds, mammals	Populations, species, functional groups	Local	High
Pest control	Invertebrate parasitoids and predators and vertebrate predators	Populations, species, functional groups	Local	High
Purification of water	Vegetation, soil micro-organisms, aquatic micro-organisms, aquatic invertebrates	Populations, species, functional groups, communities, habitats	Local–regional	Medium to high
Detoxification and decomposition of wastes	Leaf litter and soil invertebrates, soil micro-organisms, aquatic micro-organisms	Populations, species, functional groups, communities, habitats	Local–regional	Medium
Soil generation and soil fertility	Leaf litter and soil invertebrates, soil micro-organisms, nitrogen-fixing plants, plant and animal production of waste products	Populations, species, functional groups	Local	Medium
Seed dispersal	Ants, birds, mammals	Populations, species, functional groups	Local	High

The second important point to note about the typology shown in Figure 6.1 is that the supporting services have a different relationship to human well-being than the other three types of service: they do not directly benefit people, but are part of the often complex mechanisms and processes that generate other services. As Banzhaf and Boyd (2005), Boyd and Banzhaf (2005, 2006) and Wallace (2007) have noted, the MA and the wider research literature are in fact extremely ambiguous about how to distinguish between the mechanisms by which services are generated (called by some ecosystem functions) and the services themselves. This situation prevails despite the many attempts to provide systematic typologies of ecosystem functions, goods and services (Binning *et al.* 2001, Daily 1997, de Groot 1992, de Groot *et al.* 2002, MA 2005).

The problem is an important one to resolve, because unless we can be clear about what a service actually is, it is difficult to say what role 'biodiversity' plays in its generation. Wallace (2007) has been one of the most recent to comment on the problems that the MA typology poses. He suggests that if we are to use the idea of ecosystem services to help us make decisions, then it is essential that we are able to classify them in ways that allow us to make comparisons and so evaluate the consequences of different management or policy strategies. The main problem with the MA typology, according to Wallace (2007, 2008), is that it confuses *ends* with *means;* that is the benefit that people actually enjoy and the mechanisms that give rise to that service. A service is something that is consumed or experienced by people. All the rest, he argues, is simply part of the ecological structures and processes that give rise to that benefit.

Service cascades

A way of representing the logic that underlies the ecosystem service paradigm and the debates that have developed around it is shown in Figure 6.2. The diagram makes a distinction between ecological structures and processes created or generated by living organisms and the benefits that people eventually derive. In the real world the links are not as simple and linear as this. However, the key point is that there is a cascade linking the two ends of a 'production chain'. The idea is best illustrated by an example.

The presence of ecological structures like woodlands and wetlands in a catchment may have the capacity (function) of slowing the passage of surface water. This function can have the potential of modifying the intensity of flooding. It is something *humans* find useful – and not a fundamental property of the ecosystem itself – which is why it is helpful to separate out this capability and call it a function. However, whether this function is regarded as a service or not depends upon whether 'flood control' is considered a benefit. People or society will value this function differently in different places at different times. Therefore in defining what the 'significant' functions of an ecosystem are and

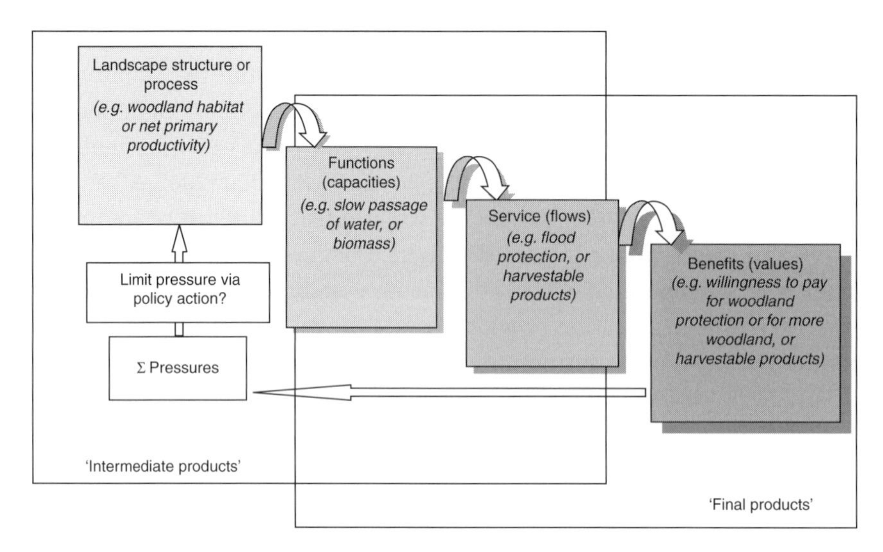

Figure 6.2 The relationship between biodiversity, ecosystem function and human well-being.

what constitutes an 'ecosystem service', an understanding of spatial context (geographical location), societal choices and values (both monetary and non-monetary) is as important as knowledge about the structure and dynamics of ecological systems themselves.

In following the cascade idea through, it is important to note the particular way that the word 'function' is being used, namely to indicate some capacity or capability of the ecosystem to do something that is potentially useful to people. This is the way commentators like de Groot *et al.* (2002) and others (e.g. Costanza *et al.* 1997, Daily 1997) use it in their account of services. However, as Jax (2005) notes, the term 'function' can mean a number of other things in ecology. It can mean something like 'capability' but it is often used more generally to refer to processes that operate within an ecosystem (like nutrient cycling or predation). This is the way Wallace (2007) uses it, although he suggests that we drop the term altogether to avoid confusion. Here, we have included the idea of functions as capabilities in Figure 6.2 to help those less familiar with the field to pick their way through current debates.

The second important idea that the cascade concept highlights is that services do not exist in isolation from people's needs. We have to be able to identify a specific benefit or beneficiary to be able to say clearly what is, or is not, a service. It is this property that led Banzhaf and Boyd (2005, p. 12) to suggest that service typologies are difficult to construct. They claim that identification of what is an ecosystem service depends on context because they are 'contingent' on 'particular human activities or wants'. The problem, which is also recognised by Wallace (2007), is illustrated by Figure 6.3, showing the different roles that 'water quality' can have in the analysis of ecosystem services and

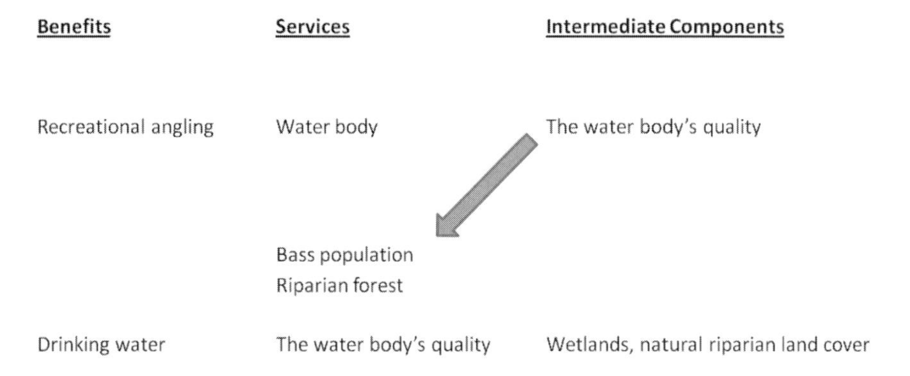

Figure 6.3 The identification of benefits, services and functions in the context of recreational angling and the provision of drinking water (after Banzhaf and Boyd 2005).

societal benefits. The quality of the water body in Figure 6.3 plays an important role in the ecosystem service 'supply chain' that produces the benefits we might recognise as 'recreational angling' and 'the provision of drinking water'. However, only in the case of drinking is the water *directly* consumed, and so only here is 'the water body's quality' to be regarded as a service. Wetlands and natural riparian land cover are important assets that help deliver that service, but they are not, according to Banzhaf and Boyd (2005), services in themselves. By contrast, for recreational angling the water body's quality is no longer the service. Here, the elements used directly are the fish population (bass) and elements of the environment, such as the presence of the surrounding vegetation which may influence the quality of the angling experience. The value of the water body's quality is taken account of in the service represented by the fish stock. In this situation the quality of the water is more a function or capability of the ecosystem; it is needed to produce the service. Notice also in Banzhaf and Boyd's scheme that services and benefits are quite distinct. As Fisher and Turner (2008) note, a benefit is something that directly impacts on the welfare of people, such as more or better drinking water or a more satisfying fishing trip. For them, in contradistinction to the definition given by the MA, a service is not a benefit – but something that changes the level of well-being (welfare).

Evolving service typologies

The message that emerges from the discussion above is that while the idea of ecosystems producing services may be attractive to the ecosystem ecology community, this is a new and developing field where concepts evolve rapidly. Nevertheless, it is clear that ecosystem services are defined by human activities and needs, an observation which has the following implications:

- The contingent nature of services suggests that it is unlikely that we can ever devise any simple, generic checklist of services that ecosystems or regions might support. Rather, lists of services like those provided by the MA should be treated more as a menu of *service–benefit themes*, within particular contexts. Concepts like 'processes', 'functions', 'services' and 'benefits' should be seen more as prompts to help sort out the complexities of a given problem rather than as a set of watertight definitions that ecosystems have to be squeezed into.

- While it is important to identify the 'final product' consumed or used, so that we can value or look at the adequacy of different levels of service output, we should not overlook the importance of the other ecosystem components on which that product depends. In fact, as Fisher and Turner (2008) and Costanza (2008) have argued, services do not have to be utilised directly by people. These authors prefer to think of intermediate and final services or products, rather than becoming trapped in arguments about what is and is not a true service (see Figure 6.2). This is a helpful perpsective, because in many cases the contribution that biodiversity makes to well-being is only part of a much larger system that may include social and economic elements.

It is likely that typologies of ecosystem services will continue to evolve and, as Costanza (2008) has pointed out, other ways of categorising are likely to emerge in addition to the type of listing suggested by the MA or Wallace (2007). For example, Costanza (2008) suggests that ecosystem services can also be classified according to their *spatial* characteristics (Table 6.2). Some, like carbon sequestration, are global in nature; since the atmosphere is so well mixed all localities where carbon is fixed are potentially useful. By contrast, others, like waste treatment and pollination, depend on proximity. 'Local proximal' services are, according to Costanza, dependent on the co-location of the ecosystem providing the service and the people who receive the benefit. He also distinguishes services that 'flow' from the point of production to the point of use (like flood regulation) and those that are enjoyed at the point at which they originate ('*in situ*' services). Finally he identifies services like cultural and aesthetic ones, which depend on the movement of users to specific places.

Costanza (2008) emphasises the need for different classification schemes, highlighting classifications that try to describe the degree to which users can be excluded from accessing services, or the extent to which users may interfere with each other when they enjoy the service (Table 6.3). Those goods and services that are privately owned or sold on a market are classified as 'excludable'. The owner or provider can regulate access to the service, normally via price. Moreover, with such services, consumers are often 'rivals' in that if one consumes or enjoys the goods the other cannot because the service or goods are finite. Most provisioning services fall into this category. A variation on

Table 6.2. *Ecosystem services classified by their spatial characteristics (after Costanza 2008).*

Global non-proximal (does not depend on proximity)
• *Climate regulation*
• *Carbon sequestration*
• *Carbon storage*
• *Cultural/existence value*
Local proximal (depends on proximity)
• *Disturbance regulation/storm protection*
• *Waste treatment*
• *Pollination*
• *Biological control*
• *Habitat/refugia*
Directional flow related: flow from point of production to point of use
• *Water regulation/flood protection*
• *Water supply*
• *Sediment regulation/erosion control*
• *Nutrient regulation*
In situ (point of use)
• *Soil formation*
• *Food production/non-timber forest products*
• *Raw materials*
User movement related: flow of people to unique natural features
• *Genetic resources*
• *Recreation potential*
• *Cultural/aesthetic*

this type of service is something like 'observing wildlife', which is in principle excludable but non-rival; what one person observes does not prevent others from experiencing the same thing. The problem with many ecosystem services, which illustrates the significance of this type of classification for ecosystem managers, is that some services are open access or 'common-pool' resources, from which it is very difficult to exclude potential users. While users may or may not interfere with each other in using those services, on the whole it is very difficult to quantify their value to society or have these values included in decision making. As Hardin (1968) pointed out many years ago, the fate of such common-pool resources is often one of progressive degradation or loss. Marine fisheries are examples of rival, non-excludable services. Many of the regulating services, like flood protection, are open access but non-rival.

A key theme of the Ecosystem Approach is the emphasis it gives to holistic thinking. If ecologists are to engage effectively in such work then they must connect with other disciplines to understand how they also look at the world (Jones and Paramor, this volume). Although ecologists and natural resource

Table 6.3. *Ecosystem services classified according to their excludability and rivalness (after Costanza 2008).*

	Excludable	Non-excludable
Rival	Rival market goods and services (most provisioning services)	Open access resources (some provisioning services)
Non-rival	Non-rival club goods (some recreation services)	Public goods and services (most regulatory and cultural services)

managers have been actively involved in the debate about ecosystem services, it is important to note that the way the concepts and terminology are developing is also being shaped by geographers, economists and a range of other social and natural scientists. Many disciplines are interested in the problems that arise at the interface of people and the environment. If we are to discover and describe fully the importance of biodiversity to human well-being then we have to understand just how the connections to well-being are made. In the last section of this chapter we will therefore look at what progress has been made in understanding the role of biodiversity in the production of ecosystem services.

Biodiversity, ecosystem function and service output

The assumption that ecosystem service output is sensitive to changes in biodiversity is implicit in many of the arguments made for conserving and restoring ecological systems. Here, we critically examine that proposition.

Schwartz *et al.* (2000) take stock of the evidence linking biodiversity and ecosystem function over the previous decade, and in particular the implications it has for the conservation debate. The review is a useful starting point, because these authors set out very clearly the kinds of issues experimental and observational studies face in resolving these key questions. They suggest that in order to use the link between biodiversity and ecosystem function as the basis for arguing that the conservation of biodiversity is important, two conditions need to be met. Crucially, we would need to show that the maintenance of ecosystem function and the output of ecosystem services are dependent on a *wide range of native species.* They also note that while a number of different types of relationship between biodiversity and ecosystem function are possible, for the conservation case to be strengthened a *direct and positive association* needs to be observed.

Figure 6.4 illustrates the kinds of relationship between biodiversity and ecosystem function that might exist. Curves A and B are those suggested by Schwartz *et al.* (2000). We have added a third relationship to those they suggested, which we will discuss later; for the moment let us consider only A and B.

The important difference between curves A and B is that in A, ecosystem function is highly sensitive to variations in biodiversity, and in B, there is a saturation

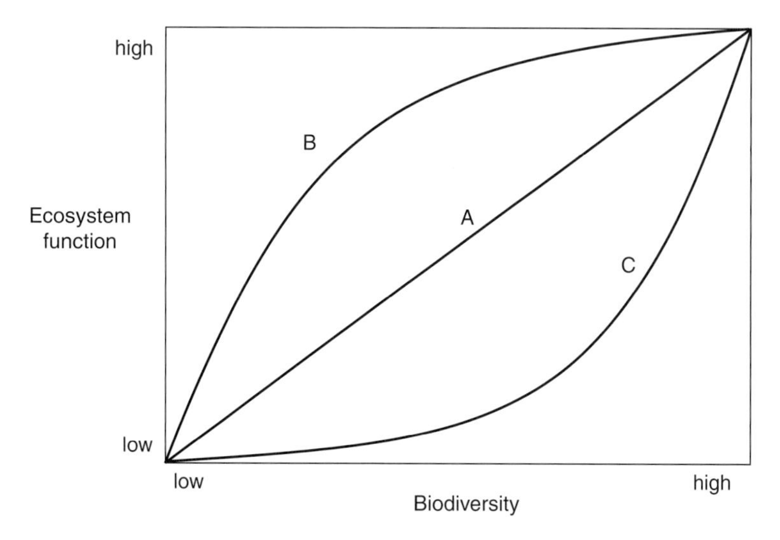

Figure 6.4 Potential relationships between biodiversity and ecosystem functioning (after Schwartz *et al.* 2000, and Kremen 2005).

effect, so that decline in ecosystem function occurs much more rapidly at low levels of species richness. Schwartz *et al.* note that the difficulty of observing relationships like curve B for the advocates of conservation is that it suggests that systems can lose much of their diversity without significantly affecting their function (operation) and potentially the benefits they provide for people. In these situations we appear to be buffered from the effects of biodiversity loss.

From their review of a range of empirical studies and modelling exercises, Schwartz *et al.* concluded that few studies supported the hypothesis that there was a simple, direct linear relationship between species richness and some measure of ecosystem function like productivity, biomass, nutrient cycling, carbon flux or nitrogen use. Instead the evidence available to them suggested that these functions did not increase proportionally above a threshold that represented a fairly low proportion of the local species pool. Others who have questioned the existence of a relationship include Aarssen (1997), Grime (1997), Huston (1997) and Wardle *et al.* (1997). Some have even suggested that any observed positive association is an artefact or sampling effect: by considering a greater number of species one is more likely to include highly productive ones (Huston and McBride 2002, Thompson *et al.* 2005).

In examining these arguments, it is important to note that there is considerable disagreement about what the evidence shows because the problem is so complex. Loreau *et al.* (2001) have, for example, suggested that any simple resolution of the question is difficult because there is considerable uncertainty about how results 'scale up' to whole landscapes and regions, and how far one can generalise across ecosystems and processes; Swift *et al.* (2004) make a similar point in the context of agricultural systems.

A further complexity arises because of the very different ways in which 'biodiversity' is measured. Biodiversity in the sense of species richness may be important for ecosystem functioning, but so might other aspects of ecosystem structure. As Díaz *et al.* (2006) point out, biodiversity in its 'broadest sense' covers not only the number of species, but also the number, abundance and composition of genotypes, populations, functional groups, and even the richness of spatial patterns exhibited by habitat mosaics and landscapes. In addition, the non-science community may have very different mental constructs of 'biodiversity', which can include iconic non-living features of the landscape, such as castles or tractors, as well as concepts such as tranquillity and scenery (Fischer and Young 2007).

Notwithstanding the difficulty of tying down the term 'biodiversity', the evidence suggests that there is a clear and direct relationship between key aspects of ecosystem function and various measures of biodiversity besides richness, such as number of functional groups or evenness. Balvanera *et al.* (2006), for example, have recently undertaken an extensive meta-analysis of experimental studies involving the manipulation of different components of biodiversity and the assessment of the consequences for ecosystem processes. Their analysis suggests that current evidence generally supports the contention that for various measures of biodiversity there *is* a positive association with a number of different measures of ecosystem functioning. They suggest that the small number of negative relationships reported in the literature tend to be associated with studies which measured properties at the population level (individual species density, cover or biomass), rather than those which looked at community-level characteristics (e.g. density, biomass, consumption). Also, the strength of the relationship between biodiversity and the measure of ecosystem function tended to be strongest at the community rather than the whole ecosystem level. A number of mechanisms underpin the relationships observed; we will consider species complementarity and the role of functional groups. The discussion will also flag up the threats that invasions of alien species might have for the output of ecosystem services and the 'insurance value' of diverse ecological systems for human well-being.

Species complementarity

Much of the discussion about the links between components of biodiversity and ecosystem functioning has been focused on what the MA call 'supporting services' or what we have called 'intermediate products'. These are not consumed by people directly but may contribute to some final benefit. Few studies have been able to trace the complete production chain from ecological structure and processes through to human well-being. As Balvanera *et al.* (2006) note, the majority of studies have focused on the consequences of biodiversity change for ecosystem productivity, and have tended to be derived from ecosystems that are easily manipulated, such as grasslands. Nevertheless, productivity is an important ecosystem function because while it may not often be a direct service, it underpins many other kinds of output. For example, more productive woodlands may support a larger standing crop of timber and hence offer greater flood or climate-regulating services. Richmond *et al.* (2007) suggest that terrestrial net primary productivity can be used as a proxy for a number of other ecosystem services, citing Gaston (2000), who observed that the output of food, timber and fibre tends to be higher in areas with high net primary production, and that at global scales, patterns of biodiversity and the associated services generally increase with net primary production. The accumulation of biomass also has a beneficial supporting role through its contribution to soil formation and the protection of soils from erosion. This view is supported by Costanza *et al.* (2007), who have investigated the inter-dependence of net primary productivity and biodiversity at the spatial scales of eco-regions in North America. They found that over half the spatial variation in net productivity could be explained by patterns of biodiversity, if the effects of temperature and precipitation were taken into account. On the basis of the relationships they develop, the authors predict that across the temperature ranges in which most of the world's biodiversity occurs, a 1 per cent change in biodiversity would result in a 0.5 per cent change in the value of ecosystem services.

Positive diversity–productivity relationships have been observed in a number of terrestrial systems at local scales. In grassland systems in Europe, for example, there is good experimental evidence that maintaining high levels of plant species diversity increases grassland productivity. Fagan *et al.* (2008) have observed that for restored grasslands on a range of soil types across southern England, there appear to be positive effects of increased species richness on ecosystem productivity. In contrast to earlier studies which monitored systems over relatively short periods, their study covered an 8-year period. Naeem *et al.* (1995), Tilman *et al.* (1996, 1997a and 1997b) and Lawton *et al.* (1998) have also provided evidence to support the existence of a direct positive relationship, whilst Cardinale *et al.* (2007) have emphasised that the productive advantage of mixtures over monocultures appears to increase over time.

Similarly, a close association between biodiversity and ecosystem functioning is apparent in many soil ecosystems. Lavelle *et al.* (2006), for example, report many experiments that show significant enhancements of plant production in the presence of Protoctista, Nematodes and Enchytraeidae, Collembola and combinations of these organisms, as well as termites, ants and earthworms. A number of factors may be responsible for such effects, including: increased release of nutrients in the plant rhizosphere; the enhancement of mutualistic micro-organisms, mycorrhizae and N-fixing microorganisms; greater protection against pests and diseases, both above and below ground; the positive effect of microorganisms on soil physical structure; and the production of plant-growth promoters.

Hooper *et al.* (2005) extensively review the issues surrounding recent discussion, and conclude that certain combinations of species are complementary in their patterns of resource use and can increase average rates of productivity and nutrient retention. 'Complementarity' is said to exist when species have niche relationships that allow the species group to capture a wider range of resources in ways that do not interfere with each other over space or time or when inter-specific interactions between species enhance the ways they collectively capture resources compared to when they grow in isolation (Cardinale *et al.* 2007, Hooper 1998). Hooper *et al.* (2005) argue that the diversity of functional traits in the species making up a community is one of the key controls on ecosystem properties. While there is a potentially large variability across ecosystems in terms of species and functional diversity, there is clear evidence that variations in ecosystem function can 'at least in part' be explained by 'differences in species or functional *composition*' (our italics).

Similar conclusions can also be drawn for many marine systems. Worm *et al.* (2006), for example, have identified a fairly strong positive association between biodiversity and productivity in marine systems, based on their meta-analysis of published experimental data. They found that increased biodiversity of both primary producers and consumers appears to enhance the ecosystem processes examined. They identified a number of explanatory factors, including complementary resource use, positive interactions between species and increased selection of highly performing species at high diversity. Moreover, they noted that the restoration of biodiversity in marine systems was also found to substantially increase productivity.

The importance of functional groups and functional traits

While there is evidence that species richness is important for maintaining ecosystem functioning, the existence of complementary relationships between species suggests that the presence of groups of species with particular properties is also significant. As Kremen (2005) notes, although we generally understand ecosystem services to be properties of whole ecosystems or communities, the functions that support them often depend upon particular populations, species, species guilds or habitat types. Thus the analysis of functional traits, the distinguishing properties of different ecological groupings, has emerged as an important area of research into understanding how ecosystem services are generated (Díaz *et al.* 2006, Balvanera *et al.* 2006).

De Bello *et al.* (2008, p. 4) define a functional trait as 'a feature of an organism which has demonstrable links to the organism's function', that is, its role in the ecosystem or its performance. 'As such', they suggest, 'functional traits determine the organism's effects on ecosystem processes or services (effect traits) and/or its response to pressures (response traits).' Although the notion of a functional trait is most easily applied at the species level, the concept can also be extended to groups of organisms with similar attributes, all of which may possess (sometimes to different degrees) similar effects or response characteristics. Whether it be at the level of single species or some wider grouping, however, there is growing consensus that 'functional diversity', that is, the type, range and relative abundance of functional traits in a community, can have important consequences for ecosystem processes (ibid.).

For example, recent work on nutrient cycling has shown that functionally diverse systems appear to be more effective in retaining nutrients than simpler systems (Hooper and Vitousek 1997, 1998). Engelhardt and Ritchie (2001) have shown that in wetland systems, not only does increased flowering-plant diversity enhance productivity, but it also aids the retention of phosphorus in the system, thereby enhancing the water purification service. The ability of vegetation to capture and store nutrients is also widely recognised in the practice of establishing buffer strips along water courses to protect them from diffuse agricultural run-off as part of water purification measures. However, the effect may not simply be additive, but more to do with the presence of particular groups of species, their particular capabilities or functions, and their abundance in relation to the levels of nutrients in the system.

The relationship between plant diversity and the retention of soil nutrients appears to be due to direct uptake of minerals by vegetation and by the effects of plants on the dynamics of soil microbial populations (Hooper and Vitousek 1997, 1998, Niklaus *et al.* 2001). The importance of diversity in relation to nutrient cycling is, in fact, particularly strong in soil ecosystems. Brussaard *et al.* (2007), for example, report evidence to suggest that increased mycorrhizal diversity positively contributes to nutrient and, possibly, water-use efficiency. Barrios (2007) has also recently reviewed the importance of the soil biota for ecosystem services and land productivity, and notes the possible positive impacts of micro-symbionts on crop yield, as a result of increases in plant-available nutrients. This is especially due to those functional groups that contribute to fertility through biological nitrogen fixation, such as *Rhizobium*, and in the case of phosphorus through arbuscular mycorrhizal fungi (see for example, Giller *et al.* 2005 and Smith and Read 1997, cited by Barrios 2007).

From the above, it is clear that by promoting particular functional responses in one group of organisms by appropriate land management, in this case the soil biota, effects may occur elsewhere by virtue of the way other organism groups react to changed ecosystem functioning. Schimel and Gulledge (1998) have made the distinction between what they call 'narrow processes', like nitrification, which are performed by a small number of key species, and other processes, such as decomposition, which tend to be dependent upon a wider range of organisms. Narrow processes may be more susceptible to changes in biodiversity or the abundance of particular functional groups, although generalisations are difficult. In the case of nitrification, for example, this may be a narrow process but the organisms which carry it out are widespread, and so it appears to be a fairly resilient process.

In relation to the processes leading to soil formation and stabilisation it has been suggested that it is not the abundance and diversity of soil organisms that are most important, but rather their functional attributes (e.g. Swift *et al.* 2004). Thus, de Ruiter *et al.* (2005) have shown that stability of the soil ecosystem is closely linked to the relative abundance of the different functional groups of organisms. Soil macrofauna (e.g. ants, termites, and earthworms) can also play an important role in the modification of soil structure through bioturbation, the production of biogenic structures (Brussaard *et al.* 2007, Lavelle and Spain 2001), and thus have an important effect on soil water and nutrient dynamics through their impact on other soil organisms (Barrios 2007). Earthworms and macro- and micro-invertebrates can improve soil structure via burrows or casts and enhance soil fertility through partial digestion and communition of soil organic matter (Zhang *et al.* 2007).

The analysis of functional groups and their associated traits is not, of course, restricted to soil ecosystems but can be applied more generally. A particular issue that has attracted much attention in the recent literature is the vulnerability of the service provided by pollinators (Losey and Vaughan 2006, Zhang *et al.* 2007). It has been estimated that the production of over 75 per cent of the world's most important crops and 35 per cent of the food produced is dependent upon animal pollination (Klein *et al.* 2007). Bees are the dominant taxa providing crop pollination services, but birds, bats, moths, flies and other insects can also be important. Pollinator diversity is essential for sustaining this highly valued service, which Costanza *et al.* (1997) estimated at global scales to be worth about $14 per ha per year. However, as Hajjar *et al.* (2008) have argued, the loss of biodiversity in agro-ecosystems through agricultural intensification and habitat decline has adversely affected pollination systems and has caused the loss of pollinators throughout the world (Kearns *et al.* 1998, Kremen *et al.* 2002, 2004, Ricketts *et al.* 2004).

The consequences of pollinator losses for ecosystem functioning have been documented by Richards (2001), who described cases where low fruit set or the setting of seeds by crops and reduction in crop yields has been attributed to a fall in pollinator diversity. There is increasing evidence that conserving wild pollinators in habitats adjacent to agriculture improves both the level and stability of pollination, leading to increased yields and income (Klein *et al.* 2003). Indeed, several studies from Europe and America have demonstrated that the loss of natural and semi-natural habitat, such as calcareous grassland, can impact upon agricultural crop production through reduced pollination services provided by native insects such as bees (Kremen *et al.* 2004).

Despite these concerns, little was known until recently about the patterns of change and what implications the loss of pollinators might have. However, an important addition to the literature has been made by Biesmeijer *et al.* (2006), who looked at the evidence available for the parallel declines in pollinators and insect-pollinated plants in Britain and the Netherlands. They compiled almost one million records for all native bees and hoverflies that could provide evidence of changes in abundance. Their analysis, which compared the period up to 1980 with that since, found that there was evidence of declines in bee abundance in both Britain and the Netherlands, but that the pattern was more mixed for hoverflies, with declines being more dependent on location and species assemblage. In both countries, those functional groups of pollinators with the narrowest habitat requirements showed the greatest declines. Moreover, in Britain, those plants most dependent on insect pollinators (the functional group represented by obligatory out-crossing plants) were also in decline, compared to other plant groups dependent on water and wind for pollination or that were self-pollinating. Wind- and water-pollinated plants were increasing while those that were self-pollinated were broadly stable. As Biesmeijer *et al.* note, it is difficult to determine whether the decline in insect-pollinated plants precedes the loss of pollinators or vice versa, but taken together, there is strong evidence of a causal connection between local extinctions of functionally linked plant and pollinator species.

Whilst species richness per se may be important in relation to the maintenance of ecosystem functioning, the role of particular keystone species or groups with specific functional capabilities should not be overlooked. This is the basis of the additional relationship that we recommend in Figure 6.4 (Curve C), which suggests that in certain circumstances, the removal of one or a small component of biodiversity can have a disproportionately large effect on ecosystem functioning (cf. Kremen 2005). There are, in fact, many situations in which particular species have been found to play pivotal roles in maintaining ecosystem processes.

Alien vs. native species

Schwartz *et al.* (2000) argued that along with evidence for a direct relationship between biodiversity and ecosystem functioning, the conservation argument may be strengthened if it can be shown that services are also dependent on the presence of a wide range of *native* species. Complementary functional relationships between species or species groups do not normally arise by chance, but rather through co-evolutionary processes. Thus it is likely that the introduction of alien species might undermine such relationships and potentially disrupt service output.

The focus of recent discussion of the threat posed by alien species to ecosystem functioning has been on two key issues. First, the properties of ecosystems that makes them resistant to invasion. Second, the impact that aliens might have on ecosystem functioning or service output.

Balvanera *et al.* (2006) suggest that the regulation of invasive species by native flora is a service of economic importance. On the basis of their meta-analysis they suggest that when plant diversity was highest, the abundance, survival and fertility of invaders were reduced. Hooper *et al.* (2005) draw similar conclusions, suggesting that susceptibility to invasion by exotic species is strongly influenced by species composition and, under similar environmental conditions, generally decreases with increasing species richness. However, other factors, such as propagule pressure, disturbance regime and resource availability also strongly influence invasion success. Klironomos (2002) has also shown that the soil microflora may be important in controlling invasibility of communities. In an experimental study, he found that while some plants maintained low densities on 'home soils' as a result of the accumulation of species-specific pathogens, plants alien to these conditions did not, and could become invasive. Thus other factors may override the effects of species richness. Hooper *et al.* (2005) caution that by increasing species richness one may actually increase the chances of invasibility within sites, if these additions result in increased resource availability, as in the case of nitrogen-fixers, or increased opportunities for recruitment through disturbance.

There is a long history of promoting the spread of alien species, often with damaging consequences for ecosystem services. Bosch and Hewlett (1982), for example, reviewed evidence from ninety-four experimental catchments, and concluded that forests dominated by introduced coniferous trees or *Eucalyptus* spp. caused larger changes than native deciduous hardwoods on water supply following planting. Calder (2002) reports that in South Africa, reforestation with exotic species such as *Pinus* spp. and *Eucalyptus* spp. significantly increased the probability of drought by reducing water flows in the dry season. In Europe, Robinson *et al.* (2003) reported significant changes in flows at the local scale, especially in *Eucalyptus globulus* plantations in Southern Portugal. In Chile, Oyarzún and Huber (1999) showed that *Pinus radiata* and *Eucalyptus* decreased water supply during the summer period.

On the basis of such evidence, Görgens and van Wilgen (2004) have suggested that invasive plants may in some situations have a negative impact on water resources. van Wilgen *et al.* (2008), for example, make an assessment of the current and potential impacts of invasive alien plants on selected ecosystem services in South Africa. They estimate that the reduction in surface water run-off as a result of the current level of invasion was equivalent to about 7% of the national total. Most of this is from the shrublands of the fynbos and grassland biomes. The analysis suggests that the potential reductions in water supply would be significantly higher if the invasive species occupied their full potential range. Impacts on groundwater recharge would be less severe. Given the current level of invasion of alien species, they estimated that in relation to the potential number of livestock that could be supported, there was a reduction in grazing capacity of around 1%, although future impacts could be closer to 71%.

Although the introduction and spread of alien or invasive species may be problematic, their control may also pose difficulties. The recent study by Marrs *et al.* (2007) has, for example, highlighted just how important it is to understand the ability of ecosystems to retain nutrients. Their work aimed to develop management strategies for the control of bracken encroachment in semi-natural communities in the UK. They found that bracken has a much greater capacity to store C, N, P, K, Ca and Mg than the other vegetation components associated with semi-natural habitats. Consequently, when bracken control measures are applied, there is a higher risk of the nutrients being released into the environment through run-off. The authors point out that this effect poses a dilemma for policies designed to control a mid-successional invasive species for conservation purposes, and that there is 'a need to balance conservation goals against potential damage to biogeochemical structure and function' (Marrs *et al.* 2007, p. 1045). Understanding the trade-offs between the different types of benefit associated with different management strategies or policy options is one of the key concerns of the Ecosystem Approach.

The insurance value of biodiversity

A novel finding of Balvanera *et al.* (2006) was that as the number of trophic levels increased between the point where the experimental intervention was made and the measurement of effects was recorded, the change in productivity was less marked. This is an interesting finding, because it suggests that ecosystems may sometimes have the capacity to buffer the effects of disturbance at one level and prevent or minimise impacts elsewhere. Such buffering has in fact been widely recognised in the ecological literature, and has been considered in much wider debates concerning the issue of ecosystem resilience.

Kremen (2005) has pointed out that if we are to manage ecosystem services successfully, then we must understand how changes in community structure collectively affect the level and stability (resilience) of ecosystem services over space and time. Although the links between diversity and stability have long been the subject of debate in ecology (Pimm 1984, Tilman 1996), the recent attention to the role of functional groups in communities throws some light onto how resilient systems are constructed.

Walker (1995), for example, has argued that ecosystem stability, measured by the probability that all species can persist, is increased if each important functional group is made up of several ecologically equivalent species, each with different responses to environmental pressures. In this sense ecological redundancy is good because it enhances ecosystem resilience. This is not to say that functionally important groups that have only one or very few species are not a priority for conservation, because their functions could be quickly lost with species extinctions (Figure 6.4, Curve C). Nevertheless, the conservation of functional redundancy may also be an important goal, if we are not to live in an unstable world.

Baumgärtner *et al.* (2007) and Quaas and Baumgärtner (2007) have made a recent analysis of the 'insurance value' of biodiversity in the provision of ecosystem services and suggest that redundancy of functional groups is an important property securing the output of ecosystem services. However, as the review of Balvanera *et al.* (2006) suggests, the buffering effects of biodiversity may be quite specific. They found that while the buffering effects of biodiversity on nutrient retention and the susceptibility to invasive species was positive, it was not so clear for disturbances related to warming, drought or high environmental variability. In the absence of further work, they conclude that a precautionary approach to the management of biodiversity is required.

Biodiversity and social–ecological ecosystems

The Ecosystem Approach emphasises that decisions about biodiversity and ecosystem services have to be looked at in a wider, social and economic context. Thus, ecologists have to find ways of linking their insights about the way ecosystems work to a broader understanding of how people benefit from nature's services, and what can be done to help sustain and improve their well-being (see also Jones and Paramor, this volume). As a result many of our most basic concepts may need to be rethought. The notion of an ecosystem is, perhaps, one of these.

As Jax (2007) has shown, the ecosystem concept has been used in a number of different ways, and he argues that there is probably no single 'right' definition for the term (see also Raffaelli and Frid, this volume). People, he observes, have modified the idea for their different purposes. It is interesting to note that the same thing is happening in the context of the debate about ecosystem services. Among other things, the cascade framework for ecosystem services that we have presented (Figure 6.2) seeks to emphasise that as scientists we are in fact dealing with a 'coupled social–ecological system' and that if we are to understand its properties and dynamics, traditional disciplinary boundaries might need to be redrawn or dissolved (see also Jones and Paramor, this volume; Raffaelli and Frid, this volume). To what extend should societal processes be included within an ecosystem?

The notion of a social–ecological system, or SES, is one that has increasingly been used in the research literature to emphasise the 'humans-in-the-environment' perspective that the Ecosystems Approach promotes. The term SES is also used to emphasise the facts that ecological and social systems are generally both highly connected and co-evolve at a range of spatial and temporal scales (see for example Folke 2006, 2007). More particularly, Anderies *et al.*

(2004) have suggested that their structure is best understood in terms of the relationships between resources, resource users and governance systems. If we follow this logic, then in defining the nature of the units that ecologists study, we must combine our scientific understanding of the relationships between biodiversity and ecosystem functioning with insights into wider social and economic structures and processes. Development of these ideas can be seen in the recent work surrounding the concept of a 'service providing unit' (SPU).

The idea of an SPU was first introduced by Luck *et al.* (2003), who argued that instead of defining a population of organisms along geographic, demographic or genetic lines, it could also be specified in terms of the service or benefit it generates at a particular scale. For example, an SPU might comprise all those organisms contributing to the wildlife interest of a site or region, or all those organisms or habitats that have a role in water purification in a catchment. It can be seen as an ecological 'footprint' of the service. As a result of work arising out of the *Rubicode*[6] Project, Vandewalle *et al.* (2007) have shown how the idea can be linked into the concept of a social–ecological system. The framework shown in Figure 6.5 is now being used to try to understand the way different pressures and drivers impact upon social–ecological systems, and the relationships between the particular components of biodiversity that generate the service, ecosystem service providers (ESPs), and ecosystem service beneficiaries (ESBs).

Models such as those shown in Figure 6.5 will enable ecologists to develop a much richer understanding of the links between biodiversity, ecosystem services and human well-being. In particular, they will help identify the kinds of trade-offs that might have to be considered between services if different development paths are chosen. An illustration of the kind of analysis required is provided by the recent work of Steffan-Dewenter *et al.* (2007), who examined the trade-offs between income, biodiversity and ecosystem functioning during tropical rainforest conversion and agroforestry intensification. Their study considered the way that incomes changed along a gradient of increasing land use intensity associated with the gradual removal of forest canopies and the reduction of shade. It appeared that there was a doubling of farmers' incomes associated with the reduction of shade from more than 80 per cent to around 30–50 per cent. However, this was associated with only limited losses of biodiversity and ecosystem function, compared to the initial conversion of forest or the complete conversion of agroforestry systems to intensive agriculture. While farmers' incomes increased further with conversion to unshaded agricultural systems, Steffan-Dewenter *et al.* (p. 4973) suggested that low-shade agroforestry represents the 'best compromise between economic forces and ecological needs'.

[6] www.rubicode.net/rubicode/index.html (Accessed 24th July, 2008)

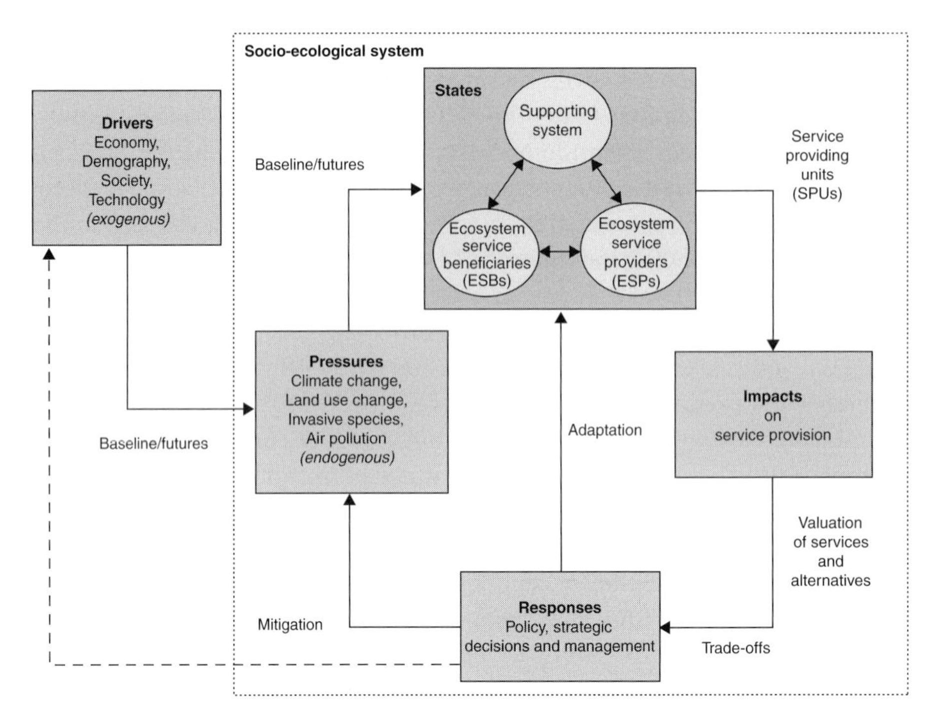

Figure 6.5 A framework for linking direct and indirect drivers, pressures and responses in a coupled socio-ecological system for assessment of the effects of environmental change drivers on ecosystem services (after Rounsevell *et al.* 2009 [in press]). Used with permission from Mark Rounsevell.

Conclusions

Ecologists will increasingly have to work alongside economists, geographers and a range of other social scientists to understand the value that biodiversity and ecosystem services have, to assess the costs and benefits of different conservation and management strategies, and to help design the new governance systems needed for sustainable development. Biodiversity has intrinsic value and should be conserved in its own right. However, the utilitarian arguments which can be made around the concept of ecosystem services and human well-being are likely to become an increasingly central focus of future debates about the need to preserve 'natural capital'. The wider research community needs to engage in such debates. Although long-term sustainable development has come to mean many things, the concept must include the maintenance of ecosystem services and the elements of human well-being that depend upon healthy ecosystems.

If the Ecosystem Approach is to be embedded in decision making then we need to understand the links between biodiversity and ecosystem services. We need to be aware of the limits of ecological functioning and how external pressures may impact on ecological structures and processes. Ecosystems can exhibit non-linear responses to such pressures and the possibility of rapid regime shifts as thresholds are crossed can mean that responses, in terms of service outputs, can be difficult to predict (Carpenter *et al.* 2006). We also need to better understand the appropriate spatial and temporal scales at which ecological systems operate if ecosystem services are to be managed wisely or restored if they have been damaged. The task of mapping ecosystem services and the construction of atlases of ecosystem services will provide the opportunity for ecologists and others to work together. It will require the development of new types of spatially explicit models that link biodiversity to ecosystem function and the benefits social–ecological systems provide in a multi-functional context. Although some progress in mapping ecosystem services has been made (see for example, Naidoo *et al.* 2008, Naidoo and Ricketts 2006, Troy and Wilson, 2006a, 2006b; and the InVEST toolbox available through the Natural Capital Project[7]) many challenges remain. These include developing better theories and better sources of data about biodiversity and the range of supporting services that living organisms provide.

The integrity of ecosystems is fundamental to human well-being. As scientists we need to understand the links between biodiversity and the benefits that people enjoy from nature. We also need to describe to the wider community how these links operate if biodiversity issues are to be taken into account in decision making. The discussion of ecosystem services is, we suggest, one way of demonstrating the relevance of the Ecosystem Approach to the needs of society.

Acknowledgement

The authors acknowledge the support of various organisations, including the Department for Environment, Food and Rural Affairs (Defra), Natural England and the European Academies Science Advisory Council (EASAC) for project work which informed the content of this review.

References

Aarssen, L.W. (1997) High productivity in grassland ecosystems: affected by species diversity or productive species? *Oikos*, **80**, 183–4.

Anderies, J.M., Janssen, M.A. and Ostrom, E. (2004) A framework to analyze the robustness of social-ecological systems from an institutional perspective. *Ecology and Society*, **9**(18) [online] available at: www.ecologyandsociety.org/vol9/iss1/art18 (Accessed 24 July 2008).

Balvanera, P., Pfisterer, A.B., Buchmann, N., He, J.-S., Nakashizuka, T., Raffaelli, D. and Schmid, B. (2006) Quantifying the evidence for biodiversity effects on ecosystem functioning and services. *Ecology Letters*, **9**, 1146–56.

Banzhaf, S., and Boyd, J. (2005) *The Architecture, and Measurement of an Ecosystem Service Index*. Discussion Paper, Resources for the Future DP 05–22.

[7] www.naturalcapitalproject.org/toolbox.html#InVEST (Accessed 24th July, 2008)

Barrios, E. (2007) Soil biota, ecosystem services and land productivity. *Ecological Economics*, **64**(2), 269–85.

Baumgärtner, S. (2007) The insurance value of biodiversity in the provision of ecosystem services. *Natural Resource Modeling*, **20**(1), 87–127.

Biesmeijer, J.C., Roberts, S.P.M., Reemer, M., Ohlemüller, R., Edwards, M., Peeters, T., Schaffers, A.P., Potts, S.G., Kleukers, R., Thomas, C.D., Settele, J. and Kunin, W.E. (2006) Parallel declines in pollinators and insect-pollinated plants in Britain and the Netherlands. *Science*, **313**, 351–4.

Binning, C., Cork, S., Parry, R. and Shelton, D. (2001) Natural assets: an inventory of ecosystem goods and services in the Goulburn Broken Catchment. CSIRO Sustainable Ecosystems, Canberra. Download; www.ecosystemservicesproject.org/html/publications/index.htm (accessed 24 August 2008).

Bosch, J.M. and Hewlett, J.D. (1982) A review of catchment experiments to determine the effect of vegetation changes on water yield and evapotranspiration. *Journal of Hydrology*, **55**, 3–23.

Boyd, J., and Banzhaf, S. (2005) Ecosystem services and government accountability: the need for a new way of judging Nature's Value. *Resources*, Summer, 16–19.

Boyd, J. and Banzhaf, S. (2006) *What are Ecosystem Services?* Discussion Paper Resources for the Future DP 06–02.

Brussaard, L., de Ruiter, P.C. and Brown, G.G. (2007) Soil biodiversity for agricultural sustainability. *Agriculture, Ecosystems & Environment*, **121**(3), 233–44.

Calder, I.R. (2002) Forests and hydrological services: reconciling public and science perceptions. *Land Use and Water Resources Research*, **2**, 2.1–2.12

Cardinale, B.J., Wright, J.P., Cadotte, M.C., Carroll, I.T., Hector, A., Srivastava, D.S., Loreau, M. and Weis, J.J. (2007) Impacts of plant diversity on biomass production increase through time because of species complementarity, *Proceedings of the National Academy Sciences*, **104**(46), 18123–8.

Carpenter, S.R., Bennett, E.M., and Peterson, G.D. (2006) Scenarios for ecosystem services: an overview. *Ecology and Society*, **11**(1), 29. [online] URL: www.ecologyandsociety.org/vol11/iss1/art29/

Chan, K.M., Pringle, R.M., Ranganathan, J., Boggs, C.L., Chan, Y.L., Ehrlich, P.R., Haff, P.K., Heller, N.E., Al-Khafaji, K. and Macmynowski, D.P. (2007) When agendas collide: human welfare and biological conservation. *Conservation Biology*, **21**(1), 51–68.

Cork, S., Shelton, D., Binning, C. and Parry, R. (2001) A framework for applying the concept of ecosystem services to natural resource management in Australia. *Third Australian Stream Management Conference*, August 27–29, 2001, Brisbane. Cooperative Research Centre for Catchment Hydrology.

Costanza, R. (2008) Ecosystem services: multiple classification systems are needed. *Biological Conservation*, **141**, 350–2.

Costanza, R., D'Arge, R., DeGroot R., Farber, S., Grasso, M., Hannon B., Limburg, K., Naeem, S., O'Neill, R., Paruelo, J., Raskin, R., Sutton, P. and van den Belt M. (1997) The value of the world's ecosystem services and natural capital. *Nature*, **387**, 253–60.

Costanza, R., Fisher, B., Mulder, K., Liu, S. and Christopher, T. (2007) Biodiversity and ecosystem services: A multi-scale empirical study of the relationship between species richness and net primary production. *Ecological Economics*, **61**, 478–91.

Daily, G.C. (1997) Introduction: what are ecosystem services? In: Daily, G.C. (Ed.), *Nature's Services: Societal Dependence on Natural Ecosystems*. Island Press, Washington, DC, 1–10.

De Bello, F., Lavorel, S., Díaz, S., Harrington, R., Bardgett, R., Berg, M., Cipriotti, P., Cornelissen, H., Feld, C., Hering, C., Martins da Silva, P., Potts, S., Sandin, L., Sousa, J.S., Storkey, J. and Wardle, D. (2008) Functional traits underlie the delivery of ecosystem services across different trophic levels. Deliverable of the Rubicode Project (download: www.rubicode.net/rubicode/outputs.html).

De Groot, R.S. (1992) *Functions of Nature: Evaluation of Nature in Environmental Planning, Management and Decision Making.* Wolters-Noordhoff, Groningen.

De Groot, R.S., Wilson, M.A. and Boumans, R.M.J. (2002) A typology for the classification, description and valuation of ecosystem functions, goods and services. *Ecological Economics*, **41**, 393–408.

de Ruiter, P.C., Neutel, A.M. and Moore, J.C. (2005). The balance between productivity and food web structure in soil ecosystems. In: Bardgett, R.D., Usher, M.B. and Hopkins, D.W. (Eds.), *Biological Diversity and Function in Soil.* Cambridge University Press, Cambridge, 139–53.

Díaz, S., Fargione, J., Chapin F.S. III and Tilman, D. (2006) Biodiversity loss threatens human well-being. *PLOS Biology*, **4**(8), e277. DOI: 10.1371/journal.pbio.0040277.

Engelhardt, K.A.M. and Ritchie, M.E. (2001) Effects of macrophyte species richness on wetland ecosystem functioning and services. *Nature*, **411**, 687–9.

European Commission (2008) The economics of ecosystems and biodiversity. Interim report. http://ec.europa.eu/environment/nature/biodiversity/economics/pdf/teeb_report.pdf (Accessed 24 July 2008).

Fagan, K.C., Pywell, R.F., Bullock, J.M. and Marrs, R.H. (2008) Do restored calcareous grasslands on former arable fields resemble ancient targets? The effect of time, methods and environment on outcomes. *Journal of Applied Ecology*, **45**(4), 1293–303.

Fischer, A. and Young, J.C. (2007) Understanding mental constructs of biodiversity: Implications for biodiversity management and conservation. *Biological Conservation*, **136**, 271–82.

Fisher, B. and Turner, K. (2008) Ecosystem services: Classification for valuation. *Biological Conservation*, **141**, 1167–9.

Folke, C. (2006) Resilience: the emerging of a perspective for socio-ecological system analysis. *Global Environmental Change*, **16**, 253–67.

Folke, C. (2007) Social-ecological systems and adaptive governance of the commons. *Ecological Research*, **22**, 14–15.

Gaston, K.J. (2000) Global patterns in biodiversity. *Nature*, **405**, 220–7.

Giller, K.E., Bignell, D., Lavelle, P., Swift, M.J., Barrios, E., Moreira, F., van Noordwijk, M., Barois, I., Karanja, N. and Huising, J. (2005) Soil biodiversity in rapidly changing tropical landscapes: scaling down and scaling up. In: Bardgett, R., Usher, M.B., Hopkins, D.W. (Eds.), *Biological Diversity and Function in Soils.* Cambridge University Press, Cambridge, 295–318.

Görgens, A.H.M. and van Wilgen, B.W. (2004) Invasive alien plants and water resources: an assessment of current understanding, predictive ability and research challenges. *South African Journal of Science*, **100**, 27–34.

Grime, J.P. (1997) Biodiversity and ecosystem function: the debate deepens. *Science*, **277**, 1260–1.

Grumbine, R.E. (1998) Seeds of ecosystem management in Leopold's *A Sand County Almanac.* Commemorative Issue Celebrating the 50th Anniversary of 'A Sand County Almanac' and the Legacy of Aldo Leopold. *Wildlife Society Bulletin*, **26**(4), 757–60.

Hajjar, R., Jarvis, D.I. and Gemmill-Herren, B. (2008) The utility of crop genetic diversity in maintaining ecosystem services. *Agriculture, Ecosystems & Environment*, **123**(4), 261–70.

Hardin, G. (1968) The tragedy of the commons. *Science*, **162**, 1243–8.

Hartje, V., Klaphake, A. and Schliep, R. (2003) *The international debate on the ecosystem approach: Critical Review, international actors, obstacles and challenges*. BfN Skripten 80.

Holdren, J.P. and Ehrlich, P.R. (1974) Human population and the global environment. *American Scientist*, **62**, 282–92.

Hooper, D.U. (1998) The role of complementarity and competition in ecosystem responses to variation in plant diversity. *Ecology*, **79**, 704–19.

Hooper, D.U., Chapin, F.S. III, Ewel, J.J., Hector, A., Inchausti, P., Lavorel, S., Lawton, J.H., Lodge, D.M., Loreau, M., Naeem, S., Schmid, B., Setälä, H., Symstad A.J., Vandermeer, J. and Wardle D.A. (2005) Effects of biodiversity on ecosystem functioning: a consensus of current knowledge. *Ecological Monographs*, **75**(1), 3–35.

Hooper, D. and Vitousek, P.M. (1997) The effects of plant composition and diversity on ecosystem processes. *Science*, **277**, 1302–5.

Hooper, D.U. and Vitousek, P.M. (1998) Effects of plant composition and diversity on nutrient cycling. *Ecological Monographs*, **68**, 121–49.

Huston, M.A. (1997) Hidden treatments in ecological experiments: re-evaluating the ecosystem function of biodiversity. *Oecologia*, **110**, 449–60.

Huston, M.A. and McBride, A.C. (2002) Evaluating the relative strengths of biotic versus abitoic controls on ecosystem processes. In: Loreau, M. Naeem, S. and Inchausti, P. (Eds.), *Biodiversity of Ecosystem Functioning: Synthesis and Perspectives*. Oxford University Press, Oxford, UK, 47–60.

Jax, K. (2005) Function and 'functioning' in ecology: what does it mean? *Oikos*, **111**(3), 641–8.

Jax, K. (2007) Can we define Ecosystems? On the confusion between definition and description of ecological concepts. *Acta Biotheoretica*, **55**(4), 341–55.

Kearns, C., Inouye, D. and Waser, N. (1998) Endangered mutualisms: the conservation of plant-pollinator interactions. *Annual Review of Ecology, Evolution and Systematics*, **29**, 83–112.

Klein, A-M., Steffan-Dewenter, I. and Tscharntke, T. (2003) Fruit set of highland coffee increases with the diversity of pollinating bees. *Proceedings of the Royal Society of London. Series B, Biological Sciences*, **270**, 955–61.

Klein, A-M., Vaissière, B.E., Cane, J.H., Steffan-Dewenter, I., Cunningham, S.A., Kremen, C. and Tscharntke, T. (2007) Importance of pollinators in changing landscapes for world crops. *Proceedings of the Royal Society of London. Series B, Biological Sciences*, **274**(1608), 303–13.

Klironomos, J.N. (2002) Feedback with soil biota contributes to plant rarity and invasiveness in communities. *Nature*, **417**, 67–9.

Kremen, C. (2005) Managing ecosystem services: what do we need to know about their ecology? *Ecological Letters*, **8**, 468–79.

Kremen C., Williams, N.M., Bugg, R.L., Fay, J.P. and Thorp, R.W. (2004) The area requirements of an ecosystem service: crop pollination by native bee communities in California. *Ecology Letters*, **7**, 1109–19.

Kremen C., Williams, N.M. and Thorp, R.W. (2002) Crop pollination from native bees at risk from agricultural intensification. *Proceedings of the National Academy of Sciences of the United States of America*, **99**, 16812–16.

Lavelle, P., Decaens, T., Aubert, M., Barot, S., Blouin, M., Bureau, F., Margerie, P., Mora,

P. and Rossi, J-P. (2006) Soil invertebrates and ecosystem services. *European Journal of Soil Biology*, **42**(Supplement 1), S3–S15.

Lavelle, P. and Spain, A.V. (2001) *Soil Ecology*. Kluwer Academic Publishers, The Netherlands.

Lawton, J.H., Bignell, D.E., Bolton, B., Bloemers, G.F., Eggleton, P., Hammond, P.M., Hodda, M., Holt, R.D., Larsen, T.B., Mawdsley, N.A. *et al.* (1998) Biodiversity inventories, indicator taxa and effects of habitat modification in tropical forest. *Nature*, **391**, 72–6.

Loreau, M., Naeem, S., Inchausti, P., Bengtsson, J., Grime, J.P., Hetcor, A., Hooper, D.U., Huston M.A., Raffaelli, D., Schmid, B., Timan, D. and Wardle, D.A. (2001) Biodiversity and ecosystem functioning: current knowledge and future challenges. *Science*, **294**, 804–8.

Losey, J.E. and Vaughan, M. (2006). The economic value of ecological services provided by insects. *Bioscience*, **56**(4), 311–23.

Luck, G.W., Daily, G.C. and Ehrlich, P.R. (2003) Population diversity and ecosystem services. *Trends in Ecology and Evolution*, **18**, 331–6.

MA [Millennium Ecosystem Assessment]. (2005) *Ecosystems and Human Well-being: Synthesis*. Island Press, Washington, DC.

Marrs, R.H., Galtress, K., Tong, C., Cox, E.S., Blackbird S.J., Heyes T.J., Pakeman, R.J. and Le Duc, M.G. (2007) Competing conservation goals, biodiversity or ecosystem services: Element losses and species recruitment in a managed moorland-bracken model system. *Journal of Environmental Management*, **85**(4), 1034–47.

Mooney, H.A. and Ehrlich, P.R. (1997) Ecosystem Services: a fragmentary history. In: Daily, G.C. (Ed.), *Nature's Services: Societal Dependence on Natural Ecosystems*. Island Press, Washington, DC, 11–19.

Naeem, S., Thompson, L.J., Lawler, S.P., Lawton, J.H. and Woodfin, R.M. (1995) Empirical evidence that declining species diversity may alter the performance of terrestrial ecosystems. *Philosophical Transactions of the Royal Society of London*, **B347**, 249–62.

Naidoo, R., Balmford, A., Costanza, R., Fisher, B., Green, B., Lehner, B., Malcolm, T.R. and Ricketts, T.H. (2008) Global mapping of ecosystem services and conservation priorities. *Proceedings of the National Academy Sciences*, **105**(28), 9495–500.

Naidoo, R. and Ricketts, T.H. (2006) Mapping the economic costs and benefits of conservation. *PLOS Biology*, **4**(11), 2153–64.

Niklaus, P.A., Kandeler, E., Leadley, P.W., Schmid, B., Tscherko, D. and Körner, C. (2001) A link between plant diversity, elevated CO_2 and soil nitrate. *Oecologia*, **127**, 540–8.

Oyarzún, C. and Huber, A. (1999) Water balance in young plantations of *Eucalyptus globulus* and *Pinus radiata* in southern Chile. *Terra*, **17**, 35–44.

Pimm, S. (1984) The complexity and stability of ecosystems. *Nature*, **307**, 321–6.

Potschin, M. and Haines-Young, R.H. (2006) Rio + 10, sustainability science and Landscape Ecology. *Landscape and Urban Planning*, **75**, 3–4, 162–74.

Quaas, M.F. and Baumgärtner, S. (2007) Natural vs. financial insurance in the management of public-good ecosystems. *Ecological Economics*, **65**, 397–406.

Richards, A. (2001) Does low biodiversity resulting from modern agricultural practice affect crop pollination and yield? *Annals of Botany*, **88**, 165–72.

Richmond, A., Kaufmann, R.K. and Mynenib, R.B. (2007) Valuing ecosystem services: A shadow price for net primary production. *Ecological Economics*, **64**, 454–62.

Ricketts, T.H., Daily, G.C., Ehrlich, P.R., and Michener, C.D. (2004) Economic value of tropical forest to coffee production.

Proceedings of the National Academy Sciences of the United States of America, **101**, 12579–82.

Robinson, M., Cognard-Plancq, A.-L., Cosandey, C., David, J., Durand, P., Führer, H.-W., Hall, R., Hendriques, M.O., Marc, V., McCarthy, R., McDonnell, M., Martin, C., Nisbet, T., O'Dea, P., Rodgers, M. and Zollner, A. (2003) Studies of the impact of forests on peak flows and baseflows: a European perspective. *Forest Ecology and Management*, **186**, 1–3, 85–97.

Rounsevell, M.D.A., Dawson, T.P. and Harrison, P.A. (in press) A conceptual framework to assess the effects of environmental change on ecosystem services. Submitted to *Biodiversity and Conservation* (June 2009).

SCEP [Study of Critical Environmental Problems]. (1970) *Man's impact on the global environment*. MIT Press, Cambridge, Massachusetts.

Schimel, J.P. and Gulledge, J. (1998) Microbial community structure and global trace gases. *Global Change Biology*, **4**(7), 745–58.

Schwartz, M.W., Bringham, C., Hoeksema, J.D., Lyons, K.G., Mills, M.H. and van Mantgem, P.J. (2000) Linking biodiversity to ecosystem function: implications for conservation ecology. *Oecologia*, **122**, 297–305.

Secretariat of the Convention for Biological Diversity. (2004) *How the Convention on Biological Diversity promotes nature and human well-being*. Secretariat of the Convention on Biological Diversity with the support of the United Nations Environment Programme (UNEP) and the Government of the United Kingdom.

Smith, K.R. (2006) Public payments for environmental services from agriculture: precedents and possibilities. *American Journal of Agricultural Economics*, **88**(5), 1167–73.

Smith, S.E. and Read, D.J. (1997) *Mycorrhizal Symbiosis*, 2nd Edn. Academic Press, New York.

Steffan-Dewenter, I., Kessler, M., Barkmann, J., Boss, M.M., Buchori, D., Erasmi, S., Faust, H., Gerold, G., Glenk, K., Gradstein, S.R., Guhardja, E., Harteveld, M., Hertel, D., Hohn, P., Kappas, M., Kohler, S., Leuschner, C., Maertens, M., Marggraf, R., Migge-Kleian, S., Mogea, J., Pitopang, R., Schaefer, M., Schwarze, S., Sporn, S.G., Steingrebe, A., Tjitrosoedirdjo, S.S., Tjitrosoedirdjo, S., Twele, A., Weber, R., Woltmann, L., Zeller, M. and Tscharntke, T. (2007) Tradeoffs between income, biodiversity, and ecosystem functioning during tropical rainforest conversion and agroforestry intensification. *Proceedings of the National Academy Sciences*, **104**, (12), 4973–8.

Swallow, B., Kallesoe, M., Iftikhar, U., Van Noordwijk, M., Bracer, C., Scherr, S., Raju, K.V., Poats, S., Duraiappah, A., Ochieng, B., Mallee, H. and Rumley, R. (2007) *Compensation and Rewards for Environmental Services in the Developing World: Framing Pan-Tropical Analysis and Comparison*. ICRAF Working Paper no. 32. Nairobi: World Agroforestry Centre.

Swift, M.J., Izac, A.-M.N. and van Noordwijk, M. (2004) Biodiversity and ecosystem services in agricultural landscapes – are we asking the right questions? *Agriculture, Ecosystems and Environment*, **104**, 113–34.

Thompson, K., Askew, A.P., Grime, J.P., Dunnett, N.P. and Willis, A.J. (2005) Biodiversity, ecosystem function and plant traits in mature and immature plant communities. *Functional Ecology*, **19**(2), 355–8.

Tilman, D. (1996) Biodiversity: population versus ecosystem stability. *Ecology*, **77**, 350–63.

Tilman, D., Wedin, D. and Knops, J. (1996) Productivity and sustainability influenced by biodiversity in grassland ecosystems. *Nature*, **379**, 718–20.

Tilman, D., Knops, J., Wedin, D., Reich, P., Ritchie, M. and Siemann, E. (1997a) The influence of functional diversity and composition on ecosystem processes. *Science*, **277**, 1300–2.

Tilman, D., Lehman, C.L. and Thomson, K.T. (1997b) Plant diversity and ecosystem productivity: theoretical considerations. *Proceedings of the National Academy of Sciences of the United States of America*, **94**, 1857–61.

Troy, A. and Wilson, M.A. (2006a) Mapping ecosystem services: Practical challenges and opportunities in linking GIS and value transfer. *Ecological Economics*, **60**(2), 435–49.

Troy, A. and Wilson, M.A. (2006b) Erratum to 'Mapping ecosystem services: Practical challenges and opportunities in linking GIS and value transfer'. *Ecological Economics*, **60**(4), 435–49.

van Wilgen, B.W., Reyers, B., Le Maitre, D.C., Richardson, D.M. and Schonegevel, L. (2008) A biome-scale assessment of the impact of invasive alien plants on ecosystem services in South Africa. *Journal of Environmental Management*, **89**(4), 336–49, on-line August 2007.doi:10.1016/j.jenvman.2007.06.015 (accessed 24 August 2008).

Vandewalle, M., Sykes, M.T., Harrison, P.A., Luck, G.W., Berry, P., Bugter, R., Dawson, T.P., Feld, C.K., Harrington, R., Haslett, J.R., Hering, D., Jones, K.B., Jongman, R., Lavorel, S., Martins da Silva, P., Moora, M., Paterson, J., Rounsevell, M.D.A., Sandin, L., Settele, J., Sousa, J.P. and Zobel, M. (2007) Review paper on concepts of dynamic ecosystems and their services. www.rubicode.net/rubicode/RUBICODE_Review_on_Ecosystem_Services.pdf (Accessed 24 July 2008).

Walker, B. (1995) Conserving biological diversity through ecosystem resilience. *Conservation Biology*, **9**(4), 747–52.

Wallace, K.J. (2007) Classification of ecosystem services: problems and solutions. *Biological Conservation*, **139**, 235–46.

Wallace, K. (2008) Ecosystem services: Multiple classifications or confusion? *Biological Conservation*, **141**, 353–4.

Wardle, D.A., Zackrisson, O., Hörnberg, G. and Gallet, C. (1997) The influence of island area on ecosystem properties. *Science*, **277**, 1296–9.

Wardle, D.A., Bardgett, R.D., Klironomos, J.N., Setälä, H., van der Putten, W.H. and Wall, D.H. (2004) Ecological linkages between aboveground and belowground biota. *Science*, **304**, 1629–33.

Westman, W.E. (1977) How much are Nature's services worth? *Science*, **197**, 960–4.

Worm, B., Barbier, E.B., Beaumont, N., Emmett Duffy, J., Folke, C., Halpern, B.S., Jackson, J.B.C., Lotze, H.K., Micheli, F., Palumbi, S.R., Sala, E., Selkoe, K.A., Stachowicz, J.J. and Watson, R. (2006) Impacts of biodiversity loss on ocean ecosystem services. *Science*, **314**, 787–90.

Zhang, W., Ricketts, T.H., Kremen, C., Carney, K. and Swinton, S.M. (2007) Ecosystem services and dis-services to agriculture. *Ecological Economics*, **64**(2), 253–60.

CHAPTER SEVEN

Ecosystem ecology and environmental management

CHRISTOPHER L. J. FRID
School of Environmental Sciences, University of Liverpool
DAVID G. RAFFAELLI
Environment, University of York

Conserving ecosystems and their biodiversity must be a shared objective of industry, the conservation community and consumers. Nowhere is this more important than in agriculture that directly depends on nature. The tight agricultural markets make this even more urgent.

Julia Marton-Lefèvre, IUCN Director General.

In the last 40 years, the area of global agricultural land has grown by 10%, but in per capita terms agricultural land area has been in decline. This trend is expected to continue as land is increasingly limited and the population grows.

From WBCSD/IUCN 2008

Introduction

With the adoption of the Convention on Biological Diversity (United Nations 1992), the sustainable management and protection of biodiversity shifted from being an option to an acknowledged necessity: sustainability is now the high-level goal of environmental management policy. Sustainability implies the ability for processes and activities to be able to continue indefinitely. Within the specific context of environmental management this has been taken to mean meeting the needs of the present without compromising the ability of future generations to meet their own needs (derived from the definition of sustainable development developed by the Brundtland Commission (World Commission on Environment and Development 1987)). More recently the concept has been expanded to explicitly include three elements, sometimes referred to as the three pillars of sustainability: environmental, social and economic (Table 7.1). For many policy makers, sustaining social structures and hence economic systems is the imperative, while many scientists would argue that you cannot have a sustainable economy without sustainable use of natural systems. The key message however is that the three pillars are linked and together provide the long-term objective for environmental management.

The challenges for those charged with managing the system are determining the limits to sustainability – i.e. what are the ways and rates of use which

Ecosystem Ecology: A New Synthesis, eds. David G. Raffaelli and Christopher L. J. Frid. Published by Cambridge University Press.

Table 7.1. *Components of the three pillars of sustainability as identified by the International Union for the Conservation of Nature (adapted from WBCSD/IUCN 2008).*

Environmental	Social	Economic
Support biodiversity and services	Foster healthy populations to realise their development potential	Provide income to rural communities
Sustain productive agriculture, avoiding encroachment on natural systems	Improve livelihoods by providing ecosystem goods	Enhance agricultural value through the value chain
Manage resources well		

can be sustained – and setting in place policies to achieve those goals. The latter is a socio-political issue while the former is very much a scientific issue and determining those limits may be the greatest challenge facing ecologists in the third millennium.

There is no easy, one-size-fits-all tool for sustainable ecosystem management, although the textbooks are littered with examples of tools and approaches that have worked well for specific ecosystems (e.g. Meffe *et al.* 2002 and references therein). Ecosystem management is challenging, partly because of the complex challenges presented by managing coupled social and biophysical systems, the science of which remains at a very young stage, and partly because of the all-things-to-all-men nature of an ecosystem (Raffaelli and Frid, this volume). For these reasons, we do not offer prescriptions for ecosystem management. Instead, we highlight some of the issues and concepts which might help those who are involved with the day-to-day management of ecosystems to construct frameworks within which they can operate. Being aware of these issues should increase the chance that decisions will lead to sustainable outcomes, or at least not close down future management options. Central to this is the need to recognise that ecosystems comprise complex, highly connected ecological and social networks, such that changes brought about by management of one component are likely to affect other components in complicated and non-linear ways. As a consequence, ecosystem managers will always find themselves managing under some degree of uncertainty, due in part to an incomplete knowledge of system dynamics and in part to what are at best probabilistic predictions about future states under climate and social change. Here we rehearse the basis of some of that uncertainty.

Environmental management, biodiversity and ecosystem sustainability

Much of the management which has environmental impact is not carried out by people who understand how ecosystems work, or have an appreciation of the importance of biodiversity in the maintenance of a habitable planet and the continued delivery of the goods and services that contribute to human well-being. The Millennium Ecosystem Assessment (MA) has set out this case well for the links between biodiversity and well-being, but whilst the MEA has widespread currency amongst the academic and science communities, as well as within some of the higher levels of government and the private sector, the MEA remains virtually unknown to the majority of those responsible for the routine management of biodiversity. Indeed, for those whose focus of management responsibilities is concerned with more physical aspects of ecosystems, (e.g. mineral extraction, house building and transport infrastructure), it may not be obvious that there is a need to consider biological components at all, other than meeting legal obligations for protected habitats and species. Such thinking persists across all aspects of the economy and it is useful therefore to be reminded of why biodiversity matters, even though it may be axiomatic to ecosystem ecologists.

At the most fundamental level, without the thin film of life that is smeared across the surface of the planet human existence would simply be impossible. Without the production of oxygen by plants the conditions for the development of metazoans (including humans) would probably not have arisen. Some planetary scientists suggest that if there were no life on Earth, the temperature on the planet's surface would rise to greater than 60°C, possibly as high as 100°C, due to the greenhouse effect of the additional atmospheric carbon dioxide and methane emitted from volcanoes. This means there would be no liquid water and the planet's surface would be a mostly arid desert. Indeed, biological processes, and their associated ecological functions, give the Earth its unique planetary spectrum (Figure 7.1) and the presence of a large amount of oxygen in the atmosphere of a planet would provide strong evidence that it might host ecosystems like those on Earth.

Thus, the answer to a policy maker who asks the question 'why should we be concerned about ecosystem ecology?' is simply that without functioning ecosystems there could be no human life on Earth. But despite our clear dependence on living systems to sustain us, humans have been relentlessly careless and profligate with the Earth's resources, potentially compromising the functioning of the planet (Holmes 2006). For instance, it is estimated that humans consume over 40 per cent of the global primary production and use over 40 per cent of the total land area (Figure 7.2).

This level and pace of environmental change, as well as its consequences, have been well documented by the MEA. One of the most important consequences of this change has been to reduce the complexity and richness of ecosystems. Haines-Young and Potschin (this volume) review the evidence that this has in turn affected levels of ecosystem functioning and service delivery, but this is also an area which has long been of interest to ecologists with respect to broader issues of ecological stability (e.g. McNaughton 1977, von Bertalanffy

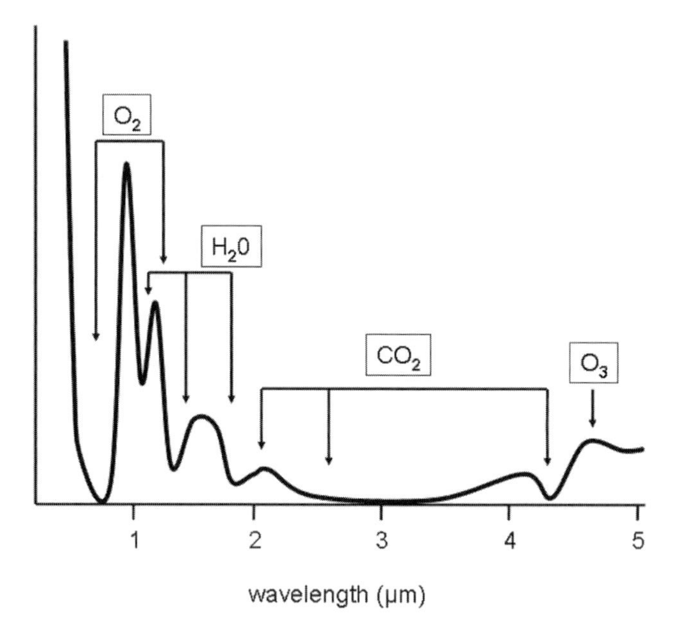

Figure 7.1 A chemical 'spectrum' of the Earth from the OMEGA spectrometer on board the Mars Express space probe. The peaks indicate that water (H_2O) and molecular oxygen (O_2) dominate, while carbon dioxide (CO_2) is also identified, as well as ozone (O_3), and several other minor constituents. (Adapted from www.astrobio.net/news/modules.php?op=modload&name=News&file=article&sid=528)

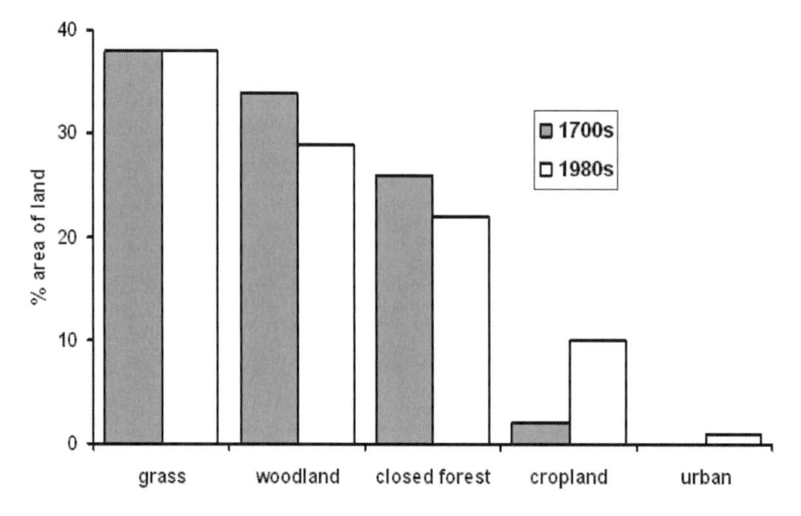

Figure 7.2 Changes in global land use from the eighteenth to twentieth centuries (data from Riebsame *et al.* 1994).

1960, McCann 2000). A logical argument can be made, often referred to as the 'Insurance Hypothesis', that species-rich systems are more resilient to perturbations. In a species-diverse system, if the abundance of one species declines, for whatever reason, other species increase to compensate, ensuring that ecosystem goods and services are maintained (von Bertalanffy 1960). Empirical modelling and experimental trials using simple mesocosm systems, e.g. bacteria and protists, lend support to the insurance hypothesis (Haines-Young and Potschin, this volume) and so provide an argument for the conservation of biodiversity and diverse systems to ensure that this insurance remains available. While this argument is both logical and supported by some experimental tests, it should be noted that many of the most extensive and productive ecosystems have a naturally low diversity, including boreal forests, lakes, bogs and estuaries (Grime 1997). In these systems there are few taxa to replace species which are lost. In addition, field-scale studies of biodiversity–ecosystem functioning relationships often contradict the results from the highly controlled simplified systems maintained in mesocosm experiments. Nevertheless, Haines-Young and Potschin (this volume) conclude that there is compelling evidence for positive relationships between ecosystem functioning and biodiversity, but it is likely that many ecosystem processes could be maintained by fewer species as long as the functional groups to which they belong remain represented (although the fewer species within each functional group, the less resilient that group is to further species loss). Thus, a priority for environmental management is to identify and then protect the key/irreplaceable species in the system, and to do so requires the development of field programmes that directly assess the functional diversity of natural functioning ecosystems (e.g. Bremner *et al.* 2003, Bremner *et al.* 2006a, 2006b, Weithoff 2003). Whilst the whole area of biodiversity–ecosystem functioning is relatively young and thus remains a little contentious, the argument that there is a link between biological diversity and the maintenance of a range of goods and services used by humankind is clearly sufficiently persuasive to have led to the development of strong policy drivers (Figure 7.3).

Recognising Darwin's Tangled Bank

The Ecosystem Approach promotes conservation and equitable, sustainable management of land, water and living resources. It relies on a scientific understanding of ecosystem structure, processes, functions and interactions (UNEP 2006). It recognises that humans, with their cultural diversity, are an integral component of many ecosystems, which along with other ecosystem elements constitute a network of interacting components: Darwin's famous metaphor of the *Tangled Bank* of species 'dependent upon each other in so complex a manner' (Darwin 1859, for the full quotation see Fenton and Spencer, this volume).

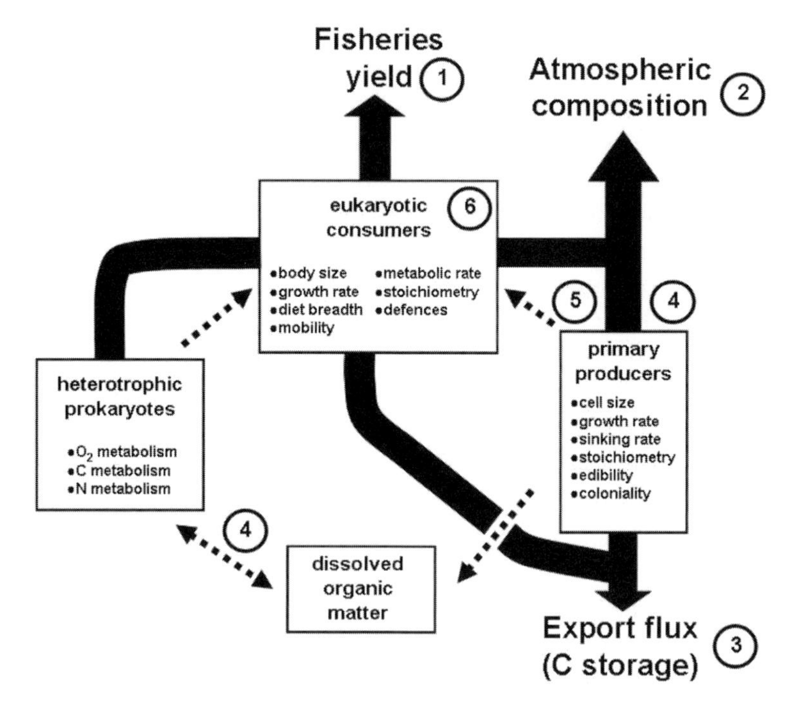

Figure 7.3 Summary of links between pelagic biodiversity and marine ecosystem processes with an indication of the areas where these impact on to environmental policy. The dotted arrows are trophic flows which link the three major compartments of the system. Traits likely to affect those trophic flows are shown within each compartment. The larger black arrows indicate how the compartments influence the delivery of ecosystem services, numbered as follows:

1. The need to ensure food supply, including protein and the health benefits of polyunsaturated fatty acids.
2. A sink for CO_2 produced by the burning of fossil fuels.
3. Climate regulation through regulation of atmospheric gases and the production of dimethyl-sulphide and hence cloud condensation nuclei which affect the planet's albedo.
4. Waste/nutrient assimilation.
5. Avoidance of toxic algal blooms.
6. Ecotourism based on naturalness and/or species of interest.

(Adapted from Duffy and Stachowicz 2006).

Changes in the state of one of the nodes in the network are likely to lead to state changes in other nodes, as perturbations spread through the system. Thus, management that focuses exclusively on a single element of an ecosystem is unlikely to lead to sustainable management, and more holistic approaches are required. A good example of wholesale ecosystem change brought about by exploitation that was focused on a single element is provided by the history of whaling in the Antarctic. Large whales were hunted almost to extinction during the last century, with the result that populations of other krill feeders, such as smaller whales, crab-eating seals and several species of penguins, increased in response to the increased availability of krill, presumed to be due to competitive release. However, cessation of whaling has not led to a recovery of whale populations, whales finding it hard to re-establish themselves in this system (Laws 1985), a phenomenon discussed at length below. Large whales are clearly embedded within a complex network of species interactions in the Antarctic, and the full consequences of their exploitation for the wider ecosystem were not appreciated at all.

The appreciation of the need for holistic approaches goes back a long way historically. For example, management in the British-controlled forests of nineteenth-century India included fire protection and maintenance of the 'natural house-hold' - what we would now refer to as the supporting ecosystem. Similarly, within the US, in addition to forestry management the early conservation movement also considered the 'natural rights' agenda. With the need to boost food and forestry production after two world wars, this holistic conservation agenda was subjugated to the provision of 'natural parks' and protection of a few high-profile species. The 1960s saw the birth of modern concepts of conservation and concern about the loss of aspects of the natural world, much of which followed from publication of Rachel Carson's *Silent Spring* (Carson 1962), and a series of species-in-crisis initiatives raised public perceptions. In general, these concerned single species. The cause of concern (hunting or pollution) was readily identified and to an extent management responses were clear and often successful. However, during the latter part of the twentieth century there were increasing numbers of cases of species that were declining to the point where they were threatened with extinction, or at least extirpation, due to degradation of their habitat, often from multiple causes. There were also many examples of single-issue species rescue plans that failed to achieve their objectives (for examples see Linquist 2008). Such cases led to the recognition of the need to protect habitats and also other components of the ecosystem such as the species' food supply. With hindsight, it seems obvious that to conserve or to exploit species sustainably, all aspects of the supporting, functioning, ecological system must also be understood and protected, but many managers today still seem to find this hard to grasp.

Sustainability and the balance of nature

The concept of environmental sustainability owes much to the idea of a 'balance of nature', a concept which is simple, intuitive to the non-specialist, but wholly without any basis in ecological science. Yet it is an amazingly persuasive paradigm that underpins much environmental policy. The idea is closely linked with the concept of 'limits to growth', first formulated by the economist and demographer Malthus and which was key in Darwin's development of the concept of natural selection (Darwin 1859). Put simply, the balance of nature implies that if a population or other component of the environment is exploited to a given level it will decline, but once that exploitation stops it will recover to its original level: the balance of nature will be re-established. If that limit is exceeded then the resource is incapable of recovery. This is the limit of sustainable use. The concept of sustainability thus implies the existence of a scientifically determinable limit value for exploitation.

The balance-of-nature perspective ignores a key aspect of the Malthus 'limits to growth' tenet: if resources are in short supply, then if harvesting removes individuals that were using that resource, other individuals will seek to gain benefit from the resources now available. These individuals may be of the harvested species, in which case they too will be removed. Therefore, in general, non-exploited taxa will ultimately benefit. In a multispecies system these adjustments could involve many taxa as competitive and predatory interactions take account of the additional 'predation' in the system coming from the harvesting. If the exploitation also alters the habitat and impacts on other species, then the system-level changes can be considerable.

The balance-of-nature model holds that once a perturbation ceases the system will swing back to its original balance. However, in some systems one or more other configurations are equally stable, so the system does not necessarily swing back to the original state, as implied above for the exploitation of whales in the Antarctic. For example, the removal of cod from the Grand Banks region off the north-east coast of North America resulted in an increase in the populations of lower-value dogfish, rays and prawns. This was probably the result of a combination of the removal of predation by large cod and release from competition with smaller cod. In the 20 years since the cod fishery was closed the stocks have failed to show a strong recovery and the system appears to have entered a new stable state as young cod seek to compete for food with the rays, dogfish and prawns and are also predated by dogfish (Rice 2002). There are those that advocate heavy fishing of the dogfish, rays and prawns to allow the cod to recover but this is potentially a high-risk strategy as there could be a third (and unknown) stable state for this system that would yield even fewer benefits. Shifts from coral-reef-dominated to algal-dominated systems have also been widely reported as examples of human-induced switching between multiple states, which have different dynamics and deliver different suites of key ecosystem services (Hughes 1994, Hughes *et al.* 2007). Raffaelli and Frid (this volume) provide other examples.

May's (1977) seminal paper on breakpoints in systems with multiple stable states used simple population models and empirical data to demonstrate not just the existence of multiple stable states in natural systems, such as grazed pasture, fish stocks and insect pest populations, but also the existence of rapid transitions (breakpoints) and of non-symmetrical trajectories (May 1977). The latter has been termed 'hysteresis', a term used in physiology, which can be defined as a system that exhibits path-dependence. That is, the current system trajectory is not just a function of the system attributes but also depends on how they have varied previously. To put this in a management context, reducing an impact may not cause the system to change back through the same intermediates as it did when the impact was applied. Hysteresis can apply not just to community composition states but also to functional dynamics (Potts *et al.* 2006).

It is clear from much of the foregoing that ecological systems can be classified as 'complex systems', the dynamics of which tend to lead to a range of behaviours that make prediction of future states difficult for managers. Even simple non-linear systems can show complex dynamics and recognition of this has had a major impact on the general public's appreciation of the limits to, for example, weather forecasting, and the suggestion that 'a butterfly flapping its wings in Florida could lead to a storm in Europe' (Gleick 1988). The potential for similarly complex dynamics in biological populations has stimulated considerable theoretical interest, but in the real world, systems with chaotic behaviour would be expected to go extinct over evolutionary time (May 1995, 1999). Thus, we may take some comfort from the fact that most natural populations will have dynamics that, while they may contain the seeds of chaos, will not normally exhibit chaotic behaviours. Of course, the concern for management and policy makers is that if human activity shifts the system outside the normal range of conditions experienced over evolutionary time, this could send the system into a region of complex dynamics making prediction and hence management extremely difficult as well as increasing the probability of wholesale system collapse.

While the occurrence and frequency of chaotic dynamics remain unclear, the widespread occurrence of multiple stable states in ecosystems is now well recognised (Beisner *et al.* 2003). As May (1977) demonstrated, this is likely to mean that those systems will also contain breakpoints or tipping-points and possible hysteresis. These have profound implications for those attempting to manage such systems (Lenton *et al.* 2008, Raffaelli and Frid, this volume, Table 7.2).

Predicting when and where these thresholds will occur (if at all) is very difficult and some authors have questioned whether knowledge of their potential existence is therefore of any use for practical management, although Groffman *et al.* (2006) offer constructive suggestions for atmospheric pollutants. Even if it turns out to be effectively impossible to identify threshold effects in advance of their manifestation, their possibility should be conveyed to stakeholders and cautious management decisions advocated to minimise their appearance in the system. At the same time, one can accept that some ecological surprises are inevitable and adjust the usage strategy for the ecosystem accordingly. For instance, the exploitation of spruce forests for timber

Table 7.2. *Climate change 'tipping elements' identified by Lenton* et al. *(2008)*

Tipping element	Assessment of current (2008) status
Arctic sea ice	Some scientists believe that the tipping point for the total loss of summer sea ice is imminent.
Greenland ice sheet	Total melting could take 300 years or more but the tipping point that could see irreversible change might occur within 50 years.
West Antarctic ice sheet	Scientists believe it could unexpectedly collapse if it slips into the sea at its warming edges.
Gulf Stream	Few scientists believe it could be switched off completely this century but its collapse is a possibility.
El Niño	The southern Pacific current may be affected by warmer seas, resulting in far-reaching climate change.
Indian monsoon	Relies on temperature difference between land and sea, which could be tipped off-balance by pollutants that cause localised cooling.
West African monsoon	In the past it has changed, causing the greening of the Sahara, but in the future it could cause droughts.
Amazon rainforest	A warmer world and further deforestation may cause a collapse of the rain supporting this ecosystem.
Boreal forests	Cold-adapted trees of Siberia and Canada are dying as temperatures rise.

in North America became increasingly optimised through focused economic investment in forestry and associated dependent industries in the last century, with more and more resources needed over time to support an increasingly unstable forest ecosystem. The large-scale collapse of such highly optimised systems seems inevitable, because they become increasingly vulnerable to perturbations (Walker and Salt 2006). Their collapse is rapidly followed by catastrophic ecological and socio-economic changes, events which helped develop the concept of adaptive cycles and resilience theory (see Raffaelli and Frid, this volume). It would be much better to manage an ecosystem by adopting resource use strategies that do not continually strive to optimise the efficiency of that usage, and instead increase the range of usages and thereby create greater resilience in both the ecological and social components of the ecosystem (ibid.).

An ecosystem-based approach

The IUCN's Commission on Ecosystem Management defines the Ecosystem Approach as a strategy for the integrated management of land, water and living resources that promotes conservation and sustainable use in an equitable way. The Ecosystem Approach places human needs at the centre of environmental management. It aims to manage human impacts on the ecosystem, based on the multiple functions that ecosystems perform and the multiple uses that are made of these functions. The Ecosystem Approach does not aim for short-term economic gains, but aims to optimise the use of an ecosystem without damaging it. It was endorsed at the fifth Conference of the Parties to the Convention on Biological Diversity (CoP 5 in Nairobi, Kenya, May 2000/ Decision V/6) as the primary framework for action under the Convention, and led to the so-called Malawi Principles (Table 7.3).

Embedded firmly within the Ecosystem Approach and the Malawi Principles is the idea that ecosystems provide services (Haines-Young and Potschin, this volume) and that there will inevitably have to be trade-offs made between different services when considering management interventions. For instance, the decision to use the environment for one activity, such as road building or the siting of complexes of offshore wind turbines, will impact on ecosystem services that were provided by the biodiversity now lost as a result of that intervention (see White *et al.*, Haines-Young and Potschin, both this volume). The full environmental costs of any management decisions therefore need to be evaluated with respect to the lost services and the net gain (or loss) of benefits presented to those in society who are the ultimate beneficiaries. This process requires knowledge and understanding of how that ecosystem works, what services it currently delivers and might potentially deliver, and how changes to one dimension (e.g. by constructing a road) affect the delivery of other services (Haines-Young and Potschin, this volume).

There are no off-the-shelf, empirical models available which will allow the manager to explore these trade-offs and their consequences, yet such information will certainly be demanded by those involved (often in conflict) in the decision-making process. Our present knowledge and understanding of the consequences of trade-offs across different spatial and temporal scales are currently poor although the subject of much research effort worldwide. For instance we are particularly ignorant of the fundamental relationships between specific biodiversity elements and specific ecosystem processes within service-providing units, such as habitats, nested within a landscape. Also, we do not understand how these different service-providing units interact with each other across the different spatial scales over which management takes place. Because of this ignorance, managers often have to make decisions on anecdotal information, logical arguments or expert judgement, running a significant risk of poor management and perhaps closing down future use options for that system.

A better evidence base is clearly desirable and the most persuasive evidence comes from large-scale, controlled experiments. For instance, if one wishes to know what the impact of forest logging is on other services such as water quality, flood protection and recreation, then one should ideally clear-fell a patch of trees and measure changes in service provision in experimental (felled) and control (intact) areas. Such experiments need to be done at ecologically relevant landscape scales, since the outcomes are likely to be scale dependent. Carrying

Table 7.3. *The twelve principles of an Ecosystem Approach to environmental management (as adopted at The Convention on Biological Diversity, Conference of the Parties 5, May 2000, Decision V/6. For fuller explanation see Annex 4 of UNEP/GPA 2006).*

1	The objectives of management of land, water and living resources are a matter of societal choice
2	Management should be decentralised to the lowest appropriate level
3	Ecosystem managers should consider the effects (actual or potential) of their activities on adjacent and other ecosystems
4	Recognising potential gains from management, there is usually a need to understand and manage the ecosystem in an economic context. Any such ecosystem-management programme should: • Reduce those market distortions that adversely affect biological diversity • Align incentives to promote biodiversity conservation and sustainable use and • Internalise costs and benefits in the given ecosystem to the extent feasible
5	Conservation of ecosystem structure and functioning, to maintain ecosystem services, should be a priority target of the Ecosystem Approach
6	Ecosystems must be managed within the limits of their functioning
7	The Ecosystem Approach should be undertaken at the appropriate spatial and temporal scales
8	Recognising the varying temporal scales and lag-effects that characterise ecosystem processes, objectives for ecosystem management should be set for the long term
9	Management must recognise that change is inevitable
10	The Ecosystem Approach should seek the appropriate balance between, and integration of, conservation and use of biological diversity
11	The Ecosystem Approach should consider all forms of relevant information, including scientific and indigenous and local knowledge, innovation and practices
12	The Ecosystem Approach should involve all relevant sectors of society and scientific disciplines

out such experiments is costly and often requires long timescales to ensure that all ramifications are manifest and thus accounted for in the decision-making process. Managers can rarely afford to wait that long and it is therefore important to estimate the required duration of the experiment before embarking on an expensive venture that may not be able to be seen through to its conclusion. This is well illustrated by the proposed use of a pathogen, calicivirus, to control rabbit populations (an alien species) in New Zealand (Figure 7.4). Like all ecosystems, the New Zealand pastureland system is complex, with rabbits embedded in a network of interactions with other species, including alien predators (cats, stoats and ferrets) and the giant skink, an endangered endemic lizard. A key uncertainty for those wishing to increase agricultural production by using the virus to remove the rabbits is the implication for skink populations: will

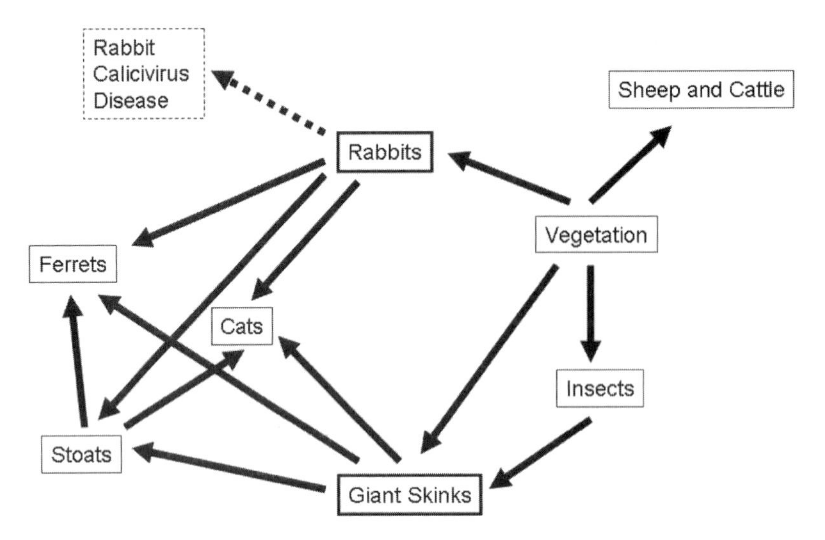

Figure 7.4 New Zealand pastureland system linkages in a putative field experiment designed to detect the effects of rabbit removal by a viral agent on the abundance of giant skink (modified from Raffaelli and Moller 2000).

they benefit or suffer? Carrying out a landscape-scale, replicated experiment to establish the wider impacts of rabbit removal would be technically challenging and expensive to set up, but, in principle, feasible. However, when Yodzis' Rule was used to estimate the time required to run the experiment in order to ensure that all outcomes would be detected (twice the sum of the generation times of the species in the longest pathway between the perturbed (rabbit) and the response (skink) species), it became apparent that the experiment would need to run for at least 50 years, an unacceptable time for a manager or policy maker (Raffaelli and Moller 2000).

A different approach to coping with this uncertainty is to construct simple box models of the system, linked by probabilistic outcomes, such as Bayesian Networks (Figure 7.5). Such models have the advantage that they require relatively little parameterisation, overcoming issues of data limitation. They also express outcomes as probabilities which reflect real uncertainties, and, as importantly, can involve all those within the decision-making process in their construction, thereby providing stakeholders with ownership of the management process and responsibility for the management outcomes, as well as informing them about the important processes within their ecosystem. Cain (2001) provides an excellent introduction to Bayesian Networks and Pollino *et al.* (2007) provide a good demonstration of its application for ecosystem management.

Weighing up the relative worth of different management, and hence service-providing, scenarios, allows losses and gains to be assessed, but this is not a straightforward process (White *et al.*, this volume). Whilst the worth of some services can be captured through monetary valuation methods, especially

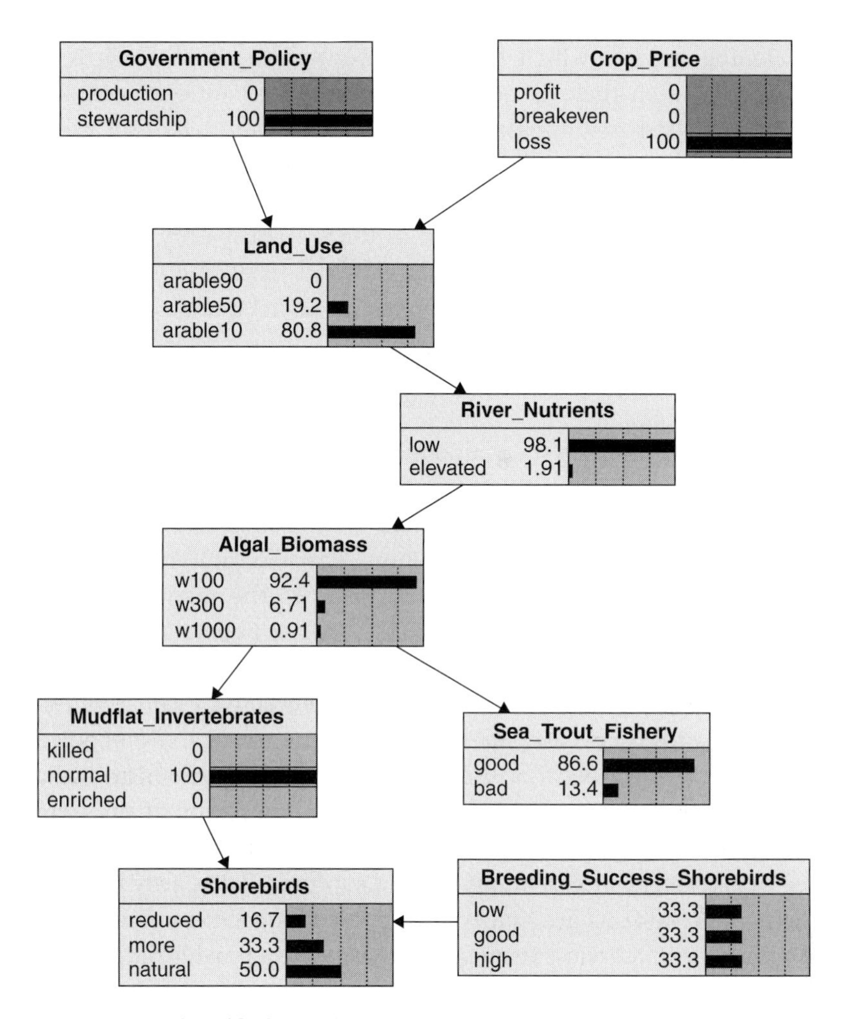

Figure 7.5 A simplified Bayesian Network for the Ythan catchment, north-east Scotland. Each box (node) presents an important factor linking land use practice in the catchment with shorebirds of conservation interest in the estuary. The probabilities shown for the factors are determined by combinations of the parent nodes and reflect expert opinion, stakeholder suggestions or empirical data. For instance, the different farming practices are determined by a combination of high-level government policy and farmers' expectations of financial returns expected from crops. Thus, when there are financial incentives for stewardship and there is a low grain price, this will encourage farmers to grow fewer hectares of arable crops. As a result, nitrogen fertiliser run-off is low, there is little growth of algal mats on mudflats, invertebrate abundance is good and shorebirds and sea trout will thrive. Altering the parent nodes to 'production' and 'profit' will lead to cascading changes in crop cover, nutrient levels, extent of algal mats, invertebrate abundance and shorebird numbers. Note that the number of shorebirds on the estuary is also driven by success on the breeding grounds many hundreds of kilometres distant (data from Raffaelli *et al.* 1999).

those provisioning services which have a market value, the real worth of other services for which there is no market are either traditionally undervalued using monetary-based approaches or cannot be captured at all using standard economic valuation methods. Combining non-commensurate measures of worth can be achieved using formal techniques, such as Multi Criteria Analysis, whilst informal approaches, such as Rural Participatory Appraisal, may be more appropriate in other contexts (Edwards-Jones *et al.* 2000). The task of accommodating a heady mix of biophysical, social and economic factors, together with sets of non-commensurate values within a decision-making framework will be one of the real challenges for an Ecosystem Approach to environmental management.

Given the increasing focus on the maintenance of the flows of services from stocks of natural and other capital which underpin the Ecosystem Approach, it seems likely that systems-based approaches, including those based on energy, will see a resurgence (Stoy, this volume, Raffaelli and Frid, this volume). Systems-based approaches have been applied to the management of many marine fisheries systems, freshwater lakes, species populations such as black bear, wetland swamps, estuaries and lagoons (reviewed in Jorgensen *et al.* 2007). A less-well-appreciated application is their potential for estimating the economic value of ecosystem components, through the concept of embodied energy or emergy. Many aspects of the environment have been historically undervalued by economists, partly due to a limited appreciation of the complex interrelationships between different ecosystem components so that the full costs of management interventions are not captured, and partly because many benefits of ecosystems are 'imperfectly owned'. A good example is the recreational and cultural services that landscape managers provide for the general public. The owner of an iconic, heather-covered grouse moor landscape in the Scottish Highlands does not charge walkers, writers, poets, artists or birdwatchers for the benefits they derive from his landscape assets, nor can he effectively deny those recreationists access to those services, since they may be perceived from far outside the owner's legal boundaries. Such services therefore generally have no market and they are consequently undervalued. Methods do exist for estimating the value of many non-market aspects of the environment, the most familiar of which are Contingent Valuation approaches, such as a person's Willingness-To-Pay. Many ecologists feel uncomfortable with such approaches because they depend heavily on consumer preferences that may change over time as knowledge and understanding and the individual's financial circumstances change. Also, those being questioned may not always be entirely honest in their answers, the process not involving an actual transaction of money.

Energy-based approaches offer an intriguing solution to these difficulties. Emergy is the energy required by nature to produce the goods or services of interest: the amount of work required to make, for example, a forest or a blue whale (Odum 1988). The more work required, the higher the value of that asset. By comparing the work done by the environment with the work of the human economy, a monetary value can be obtained for emergy, in units of emergy dollars (Em$) (Odum 1996, Odum and Odum 2000). The use of emergy in environmental accounting and decision making is described in detail in Odum (1996), but has not been widely applied, with the relatively few applications restricted to North America. Turner *et al.* (1988) applied this approach to natural landscapes in Georgia, USA. They first estimated the Gross Primary Production (GPP) of the system, a measure of how much solar energy is used to fix the carbon that, in the case of forests, provides the services of flood protection through water-flow moderation, recreation, water purification, erosion moderation, timber production and non-timber products. GPP was then converted to a relevant input into the economy, in this case fossil fuel equivalents (FFE) by considering the fuel efficiency of the natural forest resource. Finally, they converted the FFE into dollars using the ratio of Gross National Product (GNP) to total energy use in the US economy. They concluded that estimates of value based on energy analysis more properly captured the magnitude and importance of non-market aspects of biodiversity. Whilst it is clear that emergy analysis can generate far higher valuations of aspects of biodiversity than monetary-based approaches (Odum and Odum 2000), the approach does not reflect social preferences, something which is seen as an advantage by some, a disadvantage by others, and which presents a very real philosophical tension between economists and systems-analysis ecologists.

Concluding remarks

The considerable challenges in terms of both science delivery and societal understanding that accompany the adoption of an ecosystem approach mean that at present there is no simple mechanism for its delivery (Maltby *et al.* 1999). Rather a host of organisations and individuals are engaged in various aspects of translating the aspiration into practical tools for implementation of policy. This has to be done with an incomplete knowledge about the ecosystem's dynamics and the distinct possibility of major changes, like threshold effects. Armed with this knowledge, ecosystem management might seem daunting, but for most managers there is no choice. The alternative is to stand by wringing one's hands and deny a living to those whose livelihoods depend on that ecosystem, a political unreality. Probably the best that can be done is to take a pragmatic, but cautious, approach that is based on what we understand at present, which will not close down options for the future and which can be monitored so that changes in trends or approaches to limits can be identified and responded to (see White *et al.*, this volume). This is the Adaptive Management approach, where management is continually adjusted according to revealed impacts or new knowledge and understanding (Meffe *et al.* 2002). The role of the ecosystem ecologist is to provide that update of knowledge and understanding and we hope this volume will have encouraged research in this rapidly developing area.

References

Beisner, B.E., Haydon, D.T. and Cuddington, K. (2003) Alternative stable states in ecology. *Frontiers in Ecology and the Environment*, **1**(7), 376–82.

Bremner, J., Frid, C.L.J. and Rogers, S.I. (2003) Assessing marine ecosystem health: The long-term effects of fishing on functional biodiversity in North Sea benthos. *Aquatic Ecosystem Health & Management*, **6**, 131–7.

Bremner, J., Rogers, S.I. and Frid, C.L.J. (2006a) Matching biological traits to environmental conditions in marine benthic ecosystems. *Journal of Marine Systems*, **60**, 302–16.

Bremner, J., Rogers, S.I. and Frid, C.L.J. (2006b) Methods for describing ecological functioning of marine benthic assemblages using biological traits analysis (BTA). *Ecological Indicators*, **6**, 609–22.

Cain, J. (2001) *Planning improvements in natural resource management. Guidelines for using Bayesian networks to support the planning and management of development programmes in the water sector and beyond.* CEH, Wallingford.

Carson, R. (1962) *Silent Spring*. Houghton Mifflin, Boston.

Darwin, C. (1859) *On the Origin of Species by means of natural selection*. John Maurray, London.

Duffy, J.E. and Stachowicz, J.J. (2006) Why biodiversity is important to oceanography: potential roles of genetic, species, and trophic diversity in pelagic ecosystem processes. *Marine Ecology Progress Series*, **311**, 179–89.

Edwards-Jones, G., Davies B. and Hussain, S. (2000). *Ecological Economics. An introduction.* Blackell Science Ltd, Oxford.

Fraser, W.R., Trivelpiece, W.Z., Ainley, D.G. and Trivelpiece, S.G. (1992) Increases in Antarctic penguin populations: reduced competition with whales or a loss of sea ice due to environmental warming? *Polar Biology*, **11**, 525–31.

Frid, C.L.J., Paramor, O.A.L. and Scott, C.L. (2006) Ecosystem-based management of fisheries: is science limiting? *ICES Journal of Marine Science*, **63**(9), 1567–72.

Gleick, J. (1988) *Chaos*. Sphere, London.

Grime, J.P. (1997) Biodiversity and Ecosystem Function: The Debate Deepens. *Science*, **277**(5330), 1260–1.

Groffman, P.M., Baron, J.S., Blett, T., Gold, A.J., Goodman, I., Gunderson, J.H., Levinson, B.M., Palmer, M.A., Paerl, H.W., Peterson, G.D., Poff, N.L., Rejeski, D.W., Reynolds, J.F., Turner, M.G., Weathers, K.C. and Weins J. (2006) Ecological thresholds: the key to successful environmental management or an important concept with no practical application? *Ecosystems*, **9**, 1–13.

Holmes, R. (2006) Imagine Earth without people. *New Scientist*, **192**(2573), 36–41.

Hughes, T.P. (1994) Catastrophes, phase-shifts, and large-scale degradation of a Caribbean coral-reef. *Science*, **265**(5178), 1547–51.

Hughes, T.P., Rodrigues, M.J., Bellwood, D.R., Ceccarelli, D., Hoegh-Guldberg, O., McCook, L., Moltschaniwskyj, N., Pratchett, M.S., Steneck, R.S. and Willis, B. (2007) Phase shifts, herbivory, and the resilience of coral reefs to climate change. *Current Biology*, **17**(4), 360–5.

Jorgensen, S.E., Fath, B.D., Bastianoni, S., Marques, J.C., Muller, F., Nielson, S.N., Patten, B.C., Tiezzi, E. and Ulanowicz, R.E. (2007) *A New Ecology. Systems Perspective*. Elsevier, Amsterdam.

Laws, R.M. (1985) The ecology of the Southern Ocean. *American Scientist*, **73**, 26–40.

Lenton, T.M., Held, H., Kriegler, E., Hall, J.W., Lucht, W., Rahmstorf, S. and Schellnhuber, H.J. (2008) Tipping elements in the Earth's climate system. *Proceedings of the National Academy of Sciences of the United States of America*, **105**(6), 1786–93.

Linquist, S. (2008) But is it progress? On the alleged advances of conservation biology over ecology. *Biology & Philosophy*, **23**(4), 529–44.

Maltby, E., Holdgate, M., Acreman, M.C. and Weir, A. (1999) *Ecosystem management: Questions for science and society*. Royal Holloway Institute for Environmental Research, University of London, London.

May, R. (1977) Thresholds and breakpoints in ecosystems with multiple stable states. *Nature*, **269**, 471–7.

May, R.M. (1995) Necessity and chance – deterministic chaos in ecology and evolution. *Bulletin of the American Mathematical Society*, **32**(3), 291–308.

May, R. (1999) Unanswered questions in ecology. *Philosophical Transactions of the Royal Society of London. Series B-Biological Sciences*, **354**(1392), 1951–9.

McCann, K. (2000) The diversity–stability debate. *Nature*, **405**, 228–33.

McNaughton, S.J. (1977) Diversity and stability of ecological communities – comment on role of empiricism in ecology. *American Naturalist*, **111**(979), 515–25.

Meffe, G.K., Nielsen, L.A., Knight, R.L. and Schenborn, D.A. (2002) *Ecosystem management. Adaptive, community-based conservation*. Island Press, Washington, DC.

Millennium Ecosystem Assessment. (2005) *Ecosystems and human well-being: synthesis*. Island Press, Washington, DC.

Naeem, S. and Li, S.B. (1997) Biodiversity enhances ecosystem reliability. *Nature*, **390**(6659), 507–9.

Odum, H. T. (1988) Self-organisation, transformity and information. *Science*, **242**, 1132–9.

Odum, H.T. (1996) *Environmental Accounting. EMERGY and environmental decision making*. John Wiley & Sons, New York.

Odum, H.T. and Odum, E.P. (2000) The energetic basis for valuation of ecosystem services. *Ecosystems*, **3**, 21–3.

Pollino, C.A., Woodberry, O., Nicholson, A., Korb, K. and Hart, B.T. (2007) Parameterisation and evaluation of a Bayesian network for use in an ecological risk assessment. *Environmental Modelling & Software*, **22**, 1140–52.

Potts, D.L., Huxman, T., Enquist, B.J., Weltzin, J.F. and Williams, D.G. (2006) Resilience and resistance of ecosystem functional response to a precipitation pulse in a semi-arid grassland. *Journal of Ecology*, **94**(1), 23–30.

Raffaelli, D. and Moller, H. (2000) Manipulative experiments in animal ecology: do they promise more than they can deliver? *Advances in Ecological Research*, **30**, 299–338.

Raffaelli, D.G., Balls, P., Way, S., Patterson, I.J., Hohman, S.A. and Corp, N. (1999) Major changes in the ecology of the Ythan estuary, Aberdeenshire: how important are physical factors? Aquatic Conservation: Marine and Freshwater Ecosystems, **9**, 219–236.

Rice, J. (2002) Changes to the large marine ecosystem of the Newfoundland–Labrador shelf. In: Skjoldal, H.R. and Sherman, K. (Eds.) *Large Marine Ecosystems of the North Atlantic: Changing States and Sustainability*. Elsevier, Amsterdam, Netherlands, 51–103.

Riebsame, W.E., Meyer, W.B. and Turner, B.L. (1994) Modeling land-use and cover as part of global environmental-change. *Climatic Change*, **28**(1–2), 45–64.

Turner, M.G., Odum, E.P., Costanza, R. and Springer, T.M. (1988) Market and nonmarket values of the Georgia landscape. *Environmental Management*, **12**, 209–17.

UN. (2005) *Millennium Assessment. Living beyond our means: natural assets and human well-being*. UN.

UNEP/GPA. (2006) *Ecosystem-based management: Markers for assessing progress.* UNEP, The Hague.

United Nations. (1992) *Convention on Biological Diversity*. UN, New York.

Vitousek, P.M., Mooney, H.A., Lubchenco, J. and Melillo, J.M. (1997) Human domination of Earth's ecosystems. *Science*, **277**, 494–9.

von Bertalanffy, L. (1960) The theory of open systems in physics and biology. *Science*, **111**, 23–9.

Walker, B. and Salt, D. (2006) *Resilience Thinking. Sustaining ecosystems and people in a changing world*. Island Press, Washington, DC.

WBCSD/IUCN. (2008) *Agricultural Ecosystems: Facts and trends*. World Business Council for Sustainable Development, Geneva.

Weithoff, G. (2003) The concepts of 'plant functional types' and 'functional diversity' in lake phytoplankton – a new understanding of phytoplankton ecology? *Freshwater Biology*, **48**(9), 1669–75.

World Commission on Environment and Development. (1987) *Our Common Future, Report of the World Commission on Environment and Development, 1987*. UN General Assembly, Published as Annex to General Assembly document A/42/427, Development and International Co-operation: Environment, New York.

Yachi, S. and Loreau, M. (1999) Biodiversity and ecosystem productivity in a fluctuating environment: The insurance hypothesis. *Proceedings of the National Academy of Sciences of the United States of America*, **96**(4), 1463–8.

Index